"The interest in relational frame theor[...] of psychology. For anyone who wants to keep up-to-date with basic research in this area, this is the book to read."

—**Niklas Törneke MD**, author of *Learning RFT*

"Psychology is full of theories of mind, but relational frame theory (RFT) differs from all the rest in many ways. You see that when you open up this book. This lucid and engaging volume brings together the latest cutting-edge research and theory on RFT. It will challenge you in many ways, and also surprise you. It is a must-read for anyone interested in language and cognition, and especially researchers and practitioners of mindfulness and acceptance-based interventions."

—**John P. Forsyth, PhD**, professor of psychology director, Anxiety Disorders Research Program University at Albany, State University of New York

"Dymond and Roche have put together an outstanding volume that not only provides an excellent and accessible overview of relational frame theory and its rapidly accumulating empirical evidence, but also elegantly situates RFT in its proper philosophical context, makes contact with other contextually-based sciences, and elucidates nicely the many applied extensions of the theory. This book is a must and enjoyable read for anyone interested in RFT as a powerful new approach to language and cognition as well as its compelling applications."

—**Michael J. Dougher, PhD**, senior vice-provost for academic affairs, University of New Mexico

"Relational frame theory addresses the fundamental nature of symbolic thought in addition to its practical applications. It therefore deserves to be known among a large interdisciplinary audience, including my own field of evolutionary science. *Advances in Relational Frame Theory* reports on the current state of the art."

—**David Sloan Wilson**, president of the Evolution Institute and State University of New York distinguished professor of biology and anthropology, Binghamton University

ADVANCES *in* RELATIONAL FRAME THEORY

RESEARCH & APPLICATION

EDITED BY

SIMON DYMOND,
PHD, BCBA-D &

BRYAN ROCHE, PHD,
CPSYCHOL, CSCI, AFBPSS

CONTEXT PRESS
An Imprint of New Harbinger Publications, Inc.

Distributed in Canada by Raincoast Books

Copyright © 2013 by Simon Dymond and Bryan Roche
New Harbinger Publications, Inc.
5674 Shattuck Avenue
Oakland, CA 94609
www.newharbinger.com

Cover design by Amy Shoup
Acquired by Catharine Meyers

Library of Congress Cataloging in Publication Data on file

Printed in the United States of America

15   14   13

10   9   8   7   6   5   4   3   2   1

First printing

For my parents, Pat and Roger (SD)

For Suzanne, Luke, and Grace (BR)

Contents

Part I

Philosophical Foundations of Relational Frame Theory & Contextual Behavioral Science

Part II

Advances in Basic Research on Relational Frame Theory

Part III

Applications of Relational Frame Theory

List of Contributors

Clarissa Barnes, *Southern Illinois University.*

Dermot Barnes-Holmes, *National University of Ireland, Maynooth.*

Yvonne Barnes-Holmes, *National University of Ireland, Maynooth.*

Marc Bennett, *University of Leuven, Belgium.*

Michael Bordieri, *University of Mississippi.*

John T. Blackledge, *Morehead State University.*

Chad E. Drake, *Southern Illinois University*

Simon Dymond, *Swansea University.*

Mairéad Foody, *National University of Ireland, Maynooth.*

Steven C. Hayes, *University of Nevada, Reno.*

Sean Hughes, *National University of Ireland, Maynooth.*

Douglas M. Long, *University of Nevada, Reno.*

Kristen Maglieri, *Trinity College Dublin.*

Louise McHugh, *University College Dublin.*

Denis O'Hora, *National University of Ireland, Galway.*

Ruth Anne Rehfeldt, *Southern Illinois University.*

Bryan Roche, *National University of Ireland, Maynooth.*

Michael W. Schlund, *Johns Hopkins University School of Medicine; Kennedy Krieger Institute; and University of North Texas.*

Ian Stewart, *National University of Ireland, Galway.*

Triona Tammemagi, *National University of Ireland, Galway.*

Jonathan Tarbox, *Center for Autism and Related Disorders, California, United States.*

Robert Whelan, *University of Vermont.*

Kerry Whiteman, *University of Mississippi.*

Kelly G. Wilson, *University of Mississippi.*

Foreword

There are few theories in psychology that match Relational Frame Theory (RFT) in terms of depth, detail, and breadth. RFT is firmly grounded in the transparent and solid philosophical foundations of contextual behavioral science. It provides a detailed account of the nature and origins of language and thought. As such, it has important implications for a wide range of phenomena, including language development, reasoning and inference, communication and perspective taking, animal cognition, implicit cognition, developmental disorders, psychopathology, education, and organizational behavior. This book offers the first comprehensive overview of the foundations, nature, and implications of RFT. It provides both an introduction to and a state-of-the-art on RFT research.

Because the roots of RFT can be found in contextual behavioral science, the present book is bound to have the most immediate appeal to people who subscribe to this approach in psychology. However, cognitive researchers can also benefit from learning more about RFT. Being a cognitive psychologist myself, I can see at least three important ways in which RFT can contribute to cognitive science. First, it puts forward novel ideas about the development of language and thought, as well as intriguing insights in possible core differences between human and non-human cognition. These ideas are bound to inspire researchers irrespective of the research tradition in which they operate. Second, RFT highlights a number of new behavioral phenomena (e.g., different types of arbitrary applicable relational responding) that still require an explanation in terms of mediating mental mechanisms and that thus pose a challenge to cognitive researchers. Third, and perhaps most importantly, RFT provides novel ways of describing to-be-explained phenomena without using explanatory mental concepts. As such, RFT can help strengthen cognitive science by allowing researchers to conceptualize the phenomena that they study independently from the concepts that they use to explain these phenomena.

Given these strengths, it is surprising that RFT has until now received relatively little attention from cognitive researchers. This lack of interaction between behavioral and cognitive science is in part due to the fact that a somewhat different jargon is used in both approaches. Moreover, the philosophical underpinnings and aims of both approaches do not always overlap. However, unlike what is often assumed, behavioral and cognitive science can strengthen each other. Because cognitive science was founded to some extent as a reaction against the shortcomings of behavioral approaches such as those of Skinner, most cognitive researchers since then avoided any contact with behavioral psychology (and vice versa). Although avoidance behavior is often adaptive, it does interfere with the learning of changes in the environment. The proposal of RFT constituted a major change within the behavioral approach. It is time for cognitive researchers to break the habitual pattern of avoidance and to re-evaluate (contextual) behavioral science. I therefore hope that this book will not only stimulate additional RFT research but also help bridge the divide between cognitive and behavioral approaches in psychology.

Jan De Houwer
10 July 2012

Introduction

I
t is now more than 10 years since the publication of the first book on Relational Frame Theory (RFT; Hayes, Barnes-Holmes, & Roche, 2001), now affectionately known as "the purple book". That important book laid the groundwork for a progressive new behavior analysis of complex human behavior, while at the same time addressing issues of concern to psychologists from other fields. That text was radical, and even provocative, in ushering in a new post-Skinnerian era of behavioral psychology. We were both honored and invigorated to be involved with that early text and some of the earliest research into RFT.

As a now seminal text in the field, the purple book's worth should be evaluated by its fruitfulness in producing new experimental approaches to old problems and its effectiveness in predicting and influencing new and complex behavioral phenomena in the laboratory. The current volume was intended to assess just how much fruit the purple book has yielded. Ten years on, we gathered together contributions from a range of respected RFT researchers, some of whom are research "geeks" while others are working in the field making direct and immediate differences in the lives of their clients using the principles of RFT and its guiding philosophy of contextual behavioral science. Together, these researchers constitute a burgeoning, multilevel scientific community working to develop a rigorous approach adequate to the complexity of the human condition. We believe that this impressive collection of chapters is a testimony to the profound influence of the purple book, insofar as it illustrates that both the philosophy and application of RFT continue to evolve (as, indeed, they should).

As editors of the current volume, we share a set of common research interests and a history that dates back to some of the first experiments in the field of RFT supervised by our research mentor, Dermot Barnes-Holmes. Those were precious years—the kind that every enthusiastic young graduate student hopes for. It is impossible to overstate Dermot's influence on us, his assistance over the years to our various research efforts, or his role in incubating so many embryonic and even wildly ambitious ideas to maturity. Dermot has played a crucial role in the development of our research careers and in bringing RFT to the world stage. He is, and will always be, the original loon.

It is truly exciting then, that we can now present this collection to a considerably larger RFT community, but also to the wider behavior analytic community and a growing body of cognitive psychologists and associative learning theorists who have begun to see merit in the RFT approach. This impressive catalog of conceptual and empirical advances in the field attests to the vibrancy and progressiveness of the scientific approach that we have been lucky enough to know from its earliest days.

We wish to thank the authors for writing such erudite chapters and for putting up with our requests and deadlines while editing and re-editing often multiple drafts. Without them, there would simply be no book (this is clearly an instance where the whole is primary and the parts derived). We also thank the staff at New Harbinger for their enthusiastic support of the project.

Simon Dymond and Bryan Roche
July 2012

Reference

Hayes, S. C., Barnes-Holmes, D., & Roche, B. (2001). *Relational frame theory: A post-Skinnerian account of human language and cognition.* New York: Kluwer Academic.

PART I

Philosophical Foundations of Relational Frame Theory & Contextual Behavioral Science

CHAPTER ONE

Contextual Behavioral Science, Evolution, and Scientific Epistemology

Steven C. Hayes

Douglas M. Long
University of Nevada, Reno

T he purpose of this opening chapter is to examine some key philosophical assumptions of contextual behavioral science (CBS) construed as a wing of evolution science. In a series of short discussions, we will review some key aspects of contextualism and place CBS in the context of natural sciences that take a selectivist approach to behavior. Interestingly, doing so alters perspectives on derived relational learning and informs an appreciation of contextualist assumptions underlying CBS. Weaving these lines of thinking together brings greater clarity to these assumptions, but helps to demonstrate how CBS has significant implications for functional contextual perspectives throughout the natural sciences, and vice versa.

The term "contextual behavioral science" refers to *an approach to scientific system building in the behavioral sciences that is focused on the functional evolution of historically and situationally embedded actions and that extends that perspective across levels of analysis and into knowledge development itself.* As such it is a revitalization and extension of the philosophical and strategic vision of behavior analysis (e.g., Skinner, 1974) conceptualized as a functional contextual system. Within its pragmatic philosophy of science, CBS views progress as the development of scientific principles and theories which enable the *prediction and influence of the historically and situationally embedded actions of organisms with precision, scope, and depth* (Hayes, 1993).

Contextualism and Contextual Behavioral Science

As a strategy, CBS can be delineated by nine characteristics (described later) but among the most key for present purposes is its philosophical basis in "functional contextualism" (Biglan & Hayes, 1996; Hayes, 1993). A brief review is necessary. "Contextualism" is Stephen Pepper's term for the root metaphor of pragmatism in the tradition of William James (Pepper, 1942). We have long argued that behavior analysis is best understood as a contextualistic system (Hayes, Hayes, & Reese, 1988).

The unit of analysis in a contextualistic approach is the ongoing "act-in-context". Aspects of this unit can be parsed out for analytic purposes, but the whole is primary and the parts are derived. For example, consider a person who has gone to a restaurant for a lunch meeting with a friend. When he enters she gets his attention by waving her hand. The action is not merely one of raising the hand—the action is one of getting attention. If the person's arms were too tired to raise, the same functional action may have been instantiated by calling out, or standing up, or pushing back a chair. Hand raising is a participant in that whole event, but the event is not an assemblage—it is a functional whole. Without understanding the history, situation, and purpose of the act, the act itself cannot be appreciated. Getting the attention of a friend for a lunch meeting includes such contextual features as the history with this person, the circumstances that led to calling the meeting, and the agenda that will be covered.

When the act is successful—in this case when the friend sees the person—the whole event as a functional unit is satisfied. But this act is in turn nested into ever larger whole events (the need for the conversation with the friend; the ongoing relationship with the friend; and so on). For that reason, analysis in contextualism has an expansive quality as larger and larger contexts can always be considered. What prevents every single event from becoming cosmic is that analysis need only go so far that "effective action can be taken" (Skinner, 1974, p. 210). In other words, analysis itself is an "act-in-context". Stated more broadly: a contextualist scientific analysis is true if it accomplishes its purpose. This is determined by whether effective action can be taken relative to the analytic goals. Thus, the truth criterion for contextualism is said to be "successful working" (Pepper, 1942).

A pragmatic truth criterion leads to several varieties of scientific contextualism (Hayes, Hayes, Reese, & Sarbin, 1993) because different forms of contextualism can have different analytic goals. For example, descriptive contextualism seeks a personal appreciation of the participants in the whole (Hayes, 1993). Areas such as hermeneutics, feminist psychology, Marxism, or post-modernism are often contextualistic in this sense. In contrast, the analytic goal of functional contextualism is *prediction and influence with precision, scope, and depth* (Hayes, 1993). In other words, analyses are sought that simultaneously lead to the ability to predict and alter acts in context, with a limited set of principles that apply unambiguously to specific events (precision), apply to a wide range of events (scope), and cohere across scientific levels of analysis (depth).

What Is Different about Contextual Behavioral Science?

CBS focuses on *the functional evolution of historically and situationally embedded actions* in the sense that a CBS approach conceptualizes any and all actions of organisms with reference to the dynamic and ongoing functional relationship between these actions and their context, considered both in terms of their phylogenetic and ontogenetic history and the current situation. If we were to stop there, it would be fair to say that CBS is merely behavior analysis and indeed the discussion earlier about the act-in-context could be entirely redone in technical behavioral terms (we are not presenting the more general language as a technical substitute but rather as a broader introduction to the underlying philosophy). The unit we have described is the same unit as has always been emphasized in the functional wings of behavior analysis, which have considered three- and four-term contingencies to be single units, not mere collections of events. "Behavioral principles" are high precision and high scope descriptions of functional relations that describe how such units evolve.

Despite these similarities, CBS represents an expansion and elaboration of the analytic and strategic vision of behavior analysis. This is so for two primary reasons: its conceptualization of relational responding as a generalized operant, and its emphasis on scope and depth. Both of these reasons are related to how CBS extends the "act-in-context" as a unit "across levels of analysis and into knowledge development itself" as was noted in the definition of CBS above. To those two reasons, we now turn.

First, unlike traditional behavior analysis at this point in time, CBS embraces a modern functional theory of language and cognition, Relational Frame Theory (RFT; Hayes, Barnes-Holmes, & Roche, 2001). RFT modifies the functional unit of the "act-in-context" because functions can no longer be analyzed merely by looking at past experiences. For non-verbal organisms, contingencies can have a temporal extension but this occurs entirely based on past history in a direct way.

We have described it this way (Hayes, 1992, p. 113): "for a non-verbal organism time is an issue of *the past as the future in the present*. That is, based on a history of change (the 'past'), the animal is responding in the present, to present events cuing change to other events. It is not the literal future that is part of the psychology of the animal — it is the past as the future." Derived relational responding changes this perspective. It too is behavior that is established based on the individual's phylogenetic and ontogenetic history, but temporal framing interacts with direct experiences: "verbal time is *the past as the constructed future in the present*. That is, based on a history of deriving temporal sequences among events (the 'past'), the organism is responding in the present to present events by constructing a sequential relation between at least two events. Again, it is not the literal future that is part of the psychology of the verbal animal — it is the past as the constructed future" (Hayes, 1992, p. 114).

In simpler terms, RFT allows CBS to deal with verbal purpose and intention, not merely the evolution of behavioral functions in a direct way. As we will see, this implication of RFT changes the strategic vision of behavior analysis because it modifies how the verbal actions of scientists themselves are understood, demanding greater clarity about scientific assumptions and goals (Hayes, 1993) and greater flexibility in analytic practices.

Second, CBS represents an expansion and elaboration of the analytic and strategic vision of behavior analysis, and by emphasizing the levels of analysis that need to be embraced CBS places even more importance on the scope and depth of analytic accounts. This results in greater emphasis on analytic abstractive theories, and moves CBS firmly under the umbrella of approaches to evolution science that share similar analytic assumptions.

Some of the key features of the CBS approach have been described in a series of recent papers (Hayes, 2008; Levin & Hayes, 2009; Hayes, Levin, Plumb-Vilardaga, Villatte, & Pistorello, in press; Vilardaga, Hayes, Levin, & Muto, 2009). As was previously mentioned, nine key aspects of system building have been delineated. These are:

1. the explication of philosophical assumptions about science, as seen in the extensive writing on functional contextualism (e.g., Biglan & Hayes, 1996);

2. the creation of basic principles informing treatment and vice versa, as seen in RFT and its interaction with existing behavioral principles (e.g., Dougher, Hamilton, Fink, & Harrington, 2007);

3. the development of a model of pathology, intervention, and health using accessible terms tied to basic principles, as seen in the work on psychological flexibility and Acceptance and Commitment Therapy (ACT; Hayes, Strosahl, & Wilson, 2011);

4. the measurement of processes of change, as seen in measures of psychological flexibility (e.g., Bond et al., 2011);

5. the building and testing of techniques and components linked to these processes and principles, as seen in laboratory-based tests of acceptance, defusion, and values (Levin, Hildebrandt, Lillis, & Hayes, in press);

6. an emphasis on mediation and moderation in analyses of applied impact, as seen in applied studies on ACT (Hayes, Luoma, Bond, Masuda, & Lillis, 2006);

7. early testing of effectiveness, dissemination, and training strategies and developing linkage between these and the overall model (e.g., Strosahl, Hayes, Bergan, & Romano, 1998);

8. an extension of this research program across a broad range of areas and levels of analysis in an effort to detect limiting conditions (Hayes et al., 2006); and

9. the creation of a cooperative development community based on these principles (see www.contextualpsychology.org).

Because these papers on CBS and its dimensions are detailed and widely available, we will not repeat the explication of each of the dimensions *per se* in the present chapter. Rather our focus is on the implications of CBS as an expansion of the strategic vision of traditional behavior analysis due to RFT, and the nesting of CBS and RFT under evolution science, the topics to which we now turn.

CBS, RFT, and Evolution Science

Evolution Science

Evolution science approaches the structure and function of life as a matter of variation, selection, and inheritance. Although evolution science has often been viewed largely as a matter of genetic evolution, and restricted to biology, most evolution scientists have long recognized that the same basic approach applies at least to cultural evolution as well. Although the pie can be divided a number of ways, other perspectives on evolution have emerged that also include behavioral and symbolic evolution (e.g., Campbell, 1960; Jablonka & Lamb, 2005; Wilson, 2007). The defining framework of evolution offered by Campbell (1960) is sufficiently broad to serve as our starting point: blind variation and selective retention.

For behavior analysts it is not controversial to be nested under the umbrella of evolution science because Skinner formulated his approach that way. Despite the received view from outside the field, Skinner actually agreed (see Skinner, 1979, p. 209), with the seeming criticism that "the behavior of any species could not be adequately understood, predicted, or controlled without knowledge of its instinctive patterns, evolutionary history, and ecological niche" (Breland & Breland, 1961, p. 684), noting that "all behavior is due to genes, some more or less directly, the rest through the role of genes in producing the structures which are modified during the lifetime of the individual" (Skinner, 1984, p. 704). In Skinner's view, because "the behavior of organisms is a single field in which both phylogeny and ontogeny must be taken into account" (Skinner, 1977, p. 1012), "the whole story will eventually be told by the joint action of the sciences of genetics, behavior, and culture" (Skinner, 1988, p. 83).

Beyond that embrace of the biological sciences (for a complete review, see Morris, Lazo, & Smith, 2004), however, Skinner insisted that behavior be approached as a matter of variation and selection: "Selection by consequences is a causal mode found only in living things or in machines made by living things. It was first recognized in natural selection, but it also accounts for the shaping and maintenance of the behavior of the individual and evolution of cultures" (1981, p. 501).

Given that level of clarity about the need for a common selectivist approach, it is an irony of the intellectual history of the behavioral sciences that the intellectual archipelago of disparate disciplines led evolution science to largely exclude behavior analysis. This seems finally to be changing, as there is growing interest in the degree to which evolution science can contribute to the purposes of CBS and vice versa. The scope of this trend extends well beyond the specific topics in the present paper, but the discussions herein are part of such an exercise.

What Is Characteristic of Human Behavior?

The explicit goal of CBS is "the creation of a psychology more adequate to the challenge of the human condition" (Hayes et al., 2011). Although human beings are part of the natural world, a natural science of human behavior has to deal with their extraordinary characteristics. At least three are dominant: cognition, culture, and cooperation (Wilson, 2011). In relatively commonsense terms, *human cognition* refers to the capacity for symbolic thought; *culture* refers to the capacity to transmit practices and information in a cumulative fashion across the lifetimes of individuals; and *cooperation* refers to the capacity to work together to facilitate gains in large groups of individuals (Wilson, 2011). Human cognition is arguably unique, although what it means and why it may be unique is controversial. Both cooperation and culture are distinctive, but not unique. Cultural evolution has been documented in many species (Jablonka & Lamb, 2005). Shortly, we shall see how insights on these topics from evolutionary science can inform the development of RFT within the CBS agenda.

When investigating cooperation, one encounters eusocial species—species whose level of cooperation is so high that the group is arguably a new life form. Such species have evolved only a few times, and most of them are insects (Hölldobler & Wilson, 1990). Eusocial species are often spectacularly successful—ants alone compose half of the insect biomass on the planet (Hölldobler & Wilson, 1990). Human beings are arguably the only vertebrate species that can loosely be characterized as eusocial (Foster & Ratnieks, 2005), and have come to dominate the planet, perhaps because of the competitive advantages that social cooperation can provide. Perhaps because of the evident uniqueness of cognition, an analysis of human psychology has very often attempted to begin with this characteristic. However, given our unique status as eusocial vertebrates (at least as a loose characterization), it seems more sensible to assume that cooperation is fundamental and then to examine its implications for a CBS approach to cognition.

What if Cooperation Came First?

There are many reasons to think that human beings began to differentiate themselves from other primates as between-group selection began to establish high levels of cooperation and to restrain the normal tendency toward the dominance of within-group selection. Said more directly, evolution scientists have increasingly come to the conclusion that of these three characteristics—cooperation, cognition, and culture—cooperation came first (see Nowak, 2011). Kin selection and inclusive fitness accounts have long been thought to explain altruism and cooperation (e.g., Hamilton, 1964), but explanations based upon the evolutionary processes of between-group competition and multi-level selection have become more widely supported in recent years (Nowak,

Tarnita, & Wilson, 2010). In this section we examine the implications of the possibility that cooperation came first for RFT.

Although derived relational responding is viewed within RFT merely as operant behavior based on multiple exemplars, it is necessary to account both for the genetic differences that allow that training history to have its impact, and for the construction of the verbal community necessary to provide it. In the original RFT volume (Hayes et al., 2001, p. 146) we made such an attempt by pointing out that if a listener could derive a bidirectional relation from multiple exemplars, "even a single individual could have a significant behavioral advantage over others if … a weak cry of danger is 'heard' through the derivation of a bidirectional relation. Such a weak substitutive cry might not elicit running, but it might prime the animal to sense danger, to see rustling of weeds, to hear a predator, or to run just a bit more quickly if a real cry of danger is emitted. As this small difference gains prevalence in a gene pool, a group of listeners capable of deriving bidirectional relations could be created, enabling speaking that is based on bidirectional relations to be socially reinforced."

The weakness of the preceding account is that the ability to derive arbitrarily applicable relations must emerge in whole cloth in that first individual. That transition cannot be explained by the social contingencies that are appealed to in the core of the theory: no verbal community can emerge de novo to shape relational operants. If cooperation came first via fairly well understood processes of multi-level selection, cooperation itself provides a possible bridge for human language.

In non-human populations both speakers and listeners communicate in an interlocking system but "listeners acquire information from signalers who do not, in the human sense, intend to provide it" (Seyfarth & Cheney, 2003, p. 168). Speakers speak but without taking the perspective of the listener. Virtually all of the vocalizations of the great apes are a fixed set, tightly linked to emotional situations, and impossible to modify with training (for a recent review, see Tomasello, 2008). As a single example of what this means, a macaque mother seeing a predator threaten her offspring will not emit an alarm call unless she herself is threatened (Cheney & Seyfarth, 1990).

Understanding the intentions of others is not limited to people—but doing so in a cooperative context is characteristically human. For example, if a human competitor is reaching for one of several buckets that may contain a banana, great apes will happily run to the very bucket the hapless human cannot reach and retrieve the banana (Hare & Tomasello, 2004). Conversely, if the banana is hidden and a disinterested human third party *points* to the correct bucket, an ape will look at the bucket but choose it at random levels (Tomasello, Call, & Gluckman, 1997). It is as if apes understand the competitive intentions of others, but not their cooperative intentions.

Human infants use and respond to gestures in ways that seem intensely social and cooperative. For example, if while cleaning up toys together after a period of shared play an adult who has been playing with the baby points to a toy on the floor, the baby is likely to pick it up and put it in the clean-up basket; if a third party enters the room and points to the same toy, the baby will look, and may pick it up, but is unlikely to put it away (Liebal, Behne, Carpenter, & Tomasello, 2009). This fabric of cooperation can be used to explain the origin of derived relational abilities.

RFT and Cooperative Communication

Consider the simplest form of relational frame: a frame of coordination. It is the "simplest" because all derived and trained relations are the same, and because it seems to underlie the most basic linguistic act, naming. In natural language contexts, frames of coordination are taught as names are taught, but names are almost always initially taught in a context that demands that the infant be able to integrate symbols with cooperative social roles.

Take the training of a concrete noun as an example. An object is held up and an oral name is stated. Later the oral name is stated and the object is selected from an array. Initially this occurs without the child being asked to respond, but soon enough each expressive and receptive episode contains prompts and delays. For example, the object is held up and the child is asked, "What's that?" and a delay occurs, often with prompts (e.g., the first sound in the name), in order to encourage production of the name. Conversely, in the receptive episode, in the presence of the array the question is repeated with delays and subtle prompts (e.g., "Where's the ball? …. Where's the ball?" … [the ball is pushed forward slightly] "Where's the ball?").

Merely being able to produce both expressive and productive episodes is not enough. The child "understands what a word means" only when finally each role can be *derived* from the other with an entirely new name → object relation. This is the same discrimination made by RFT—framing skills are acts of derivation.

But our point here is slightly different: When teaching frames of coordination, *cooperative communication provides a model for derived relational responding itself.* What is derived in a frame of coordination can be considered to be in part the coordination of roles between a speaker and listener who are sharing a common ground of intentional cooperation. If cooperation came first, and thus the perspective of others and their roles could be included in acts of communication, it is not a big step to imagine social support for teaching that a object → sign relation from the point of view of a speaker (see a wolf → say "wolf") implies a sign → object relation from the point of view of the listener (hear "wolf" → look for wolf).

One would have to imagine that initially, despite their cooperative history, not all humans could learn to derive the mutuality of roles in a frame of coordination, nor transfer the psychological functions of objects via such derived relations and their combinations, despite training to do so. But unlike the account given in Hayes et al. (2001, p. 146), the cooperative context provides a reason for social training and a ready reason for a continuous process of genetic, behavioral, and cultural evolution. A group that had even a few speakers and listeners competent in the reversibility of roles would be advantaged in their ability to cooperate in accomplishing tasks, but only if they could respond to symbols with reversibility of roles. This provides a far more evolutionarily plausible way of creating a verbal community capable of shaping derived relational responding.

Something like this has been modeled mathematically by evolution scientists (for an overview, see Nowak, 2011). This evolutionary process could quickly create a minimally competent social/verbal community that could systematically shape additional relational frames, modeled on the perspective taking embedded in the mutuality of speaker and listener roles but no longer attached to them.

This is perhaps relatively obvious with deictic relations, but it applies as well to all instances of arbitrarily applicable derived relational responding. For example, comparative relational frames that enable the derivation of A > B given training in B < A can be thought of as learning that from the point of view of A, B is smaller, but from the point of view of B, A is larger. Said in a more colorful way, the human mind does not just enable cooperation, it is modeled on it. Human cognition is itself an act of cooperation.

The Explosion of Language

Once this process began, the enormous variability it afforded would feed the engine of behavioral and symbolic evolution. If an animal learns eight object → sign relations, there are eight behavioral events in the repertoire available to influence other behaviors. If nonarbitrary comparisons have been trained, it is possible that the eight objects can also be related non-arbitrarily one to the other (for example, if there are two types of fruit, one may be preferred). Up to 56 such relations are possible.

A human being learning the same set has more options. Each relation is mutual, so there are 16 relations available. Each object can in principle be related to all others in both directions (56 more); the same with each object (56 more); or each sign → object relation (56 more); or each object → sign relation (56 more). And then all of these can be combined into pairs, and triads, and quartets. Depending on assumptions, a simple set of eight object → sign relations can lead to tens of thousands of derived relations, and this is before considering contextual control (over relational responses and transformations of stimulus functions) that might further increase that number.

If variation and selective retention is the principle behind evolution, this kind of variation can enormously increase the ability of selection to play a role in human behavior at the symbolic and cultural level. Evolution scientists understand the adaptive implications of symbolic events (e.g., Deacon, 1997) but to date they do not yet have the concept of derived relational responses that will give more precision and even applied implications to an evolutionary account of human language and cognition.

CBS and the Divide between Evolution Science and Contextualism

Evolutionists and behaviorists alike have often adopted elemental realist assumptions, and viewed truth as a matter of the correspondence between words and events, as if this is necessary to a natural science perspective. This creates an internally contradictory linkage between epistemology and a selectivist account, however, because it ignores a natural science account of the scientist's behavior.

Post-Modernism, Realisms and Relativisms, and Evolutionary Epistemology

The beginning of the post-modern era in science can be construed as the fall of Positivist and Popperian logic as guiding principles for evaluations of progressivity (see Moore, 2010). Philosophers have since sought to fill this void with variants of realism, relativism, and pragmatism. Realist philosophers (e.g., Hacking, 1983; McMullin, 1984) have adopted the view that the principal aim of science is to develop theories that accurately represent reality. Under this assumption, successful theories are said to be instrumental as a consequence of their "genuine reference to" or "truth as correspondence with" the world. Putnam (1978) and Boyd (1983), for example, have argued that the *only* reasonable explanation for the success of scientific methodology is that our theories are gradually converging upon a true representation of the world. While this reasoning squares well with common sense, it runs into a number of logical and historical challenges.

History is replete with scientific theories and methodologies that, while enjoying both wide social acceptance and instrumental success, are nonetheless viewed today as radically misguided and false (Kuhn, 1962; Laudan, 1981). For example, navigational methods built upon the Aristotelian worldview of ancient Greek astronomy have utility even today despite being viewed as incorrect (Kuhn, 1957). Thus, genuine reference to reality is evidently neither a necessary nor a sufficient condition for empirical success. Furthermore, logical systems advanced as a means of establishing genuine reference have never mapped cleanly upon actual experimentation (see Moore, 2010). Part of the problem is that although a correspondence theory views truth in absolute terms, evidence is always partial and conditional. While some realists respond by finding solace in the notion of verisimilitude—or *closeness* to truth—this very concept shows the empirical problem. While spatial closeness is easily grasped and measured, verbal and logical closeness is little else but a loose metaphor, unless or until the absolute truth was already known so that the degree of "closeness" might be examined.

In reaction to such concerns, some in the post-modern era have taken a strongly *relativist* position, and argued that scientific knowledge has *no* special privilege over other forms of knowledge, and that apparent "progress" is *merely* socially and politically driven. Some sociologists, for example, have likened scientific methodology to the social rituals of ancient tribes (Bloor, 1976). Ironically, despite their differences, the modern traditionalist and post-modern reactionary camps actually share an essential foundational assumption (as Laudan, 1977; 1996; and Roche & Barnes-Holmes, 2003, pointed out), namely that if *scientific progress is possible, it must be contingent upon the development of theories which genuinely refer to the world.* With this assumption, it seems that one must either choose between grappling with verisimilitude or rejecting the notion of scientific progressivity altogether.

This attachment to notions of referential truth is similarly seen in the wing of philosophy of science that has attempted to apply universal Darwinian principles to verbal behavior: Evolutionary epistemology (for a review, see Radnitzky & Bartley, 1987). Although this is now a large and diverse area, its earliest manifestations (e.g., Campbell, 1959) depended both on *hypothetical realism*—that there is an external, physical world whose constituents are not existentially dependent upon their perception by humans—and on *the epistemology of the other*—that is, on the analysis of how organisms come to know, rather than also examining the knowing of the scientist. In this view, once again, language is focused on representing the world—that is, when a true statement in some way "stands for" or "refers to" a corresponding property or state of affairs in the mind-independent reality.

The Inconsistency of the Realist Agenda with a Functional Approach

Our construal of language as behavior arising naturally within the phylogenetic and ontogenetic contingencies has provocative consequences for a correspondence theory of truth. Where else in the content of the natural sciences are such notions found? When an acid and a base react to produce salt and water, one does not say that these products "refer to" their reactants. When a child is born, one does not say that the offspring is "about" its ancestors. If relational operants are merely adaptive behaviors that emerged in part because of their impact on social cooperation and effectiveness, as well as on individual adaptation, then their success is a function of individual and social utility, not of "reference"—even when relational operants take the form of scientific theories. Note well what we are doing here: applying a functional contextual perspective to knowledge claims *made by the scientist*.

Campbell (1959) is explicit in this area about what must be done for an evolutionist to hold firmly to realist assumptions and a correspondence-based view of truth: "no effort is made to justify 'my own' knowledge processes" (p. 157). In other words, elemental realist assumptions cohere with a selectivist account *provided the scientist is excluded from the account*. Campbell excuses the inconsistency on the grounds that it avoids solipsism (1959, p. 157), but this undeniable inconsistency, so far as we know, does not exist in any other area of natural science. The physicist need not exclude "doing physics" from physics to do physics; the chemist need not exclude "doing chemistry" from chemistry to do chemistry.

Campbell recognizes that evolutionary and operant processes "do not differ in any epistemological fundamental: both are blindly pragmatic and inductive" (1959, p. 160), and if that insight were to be applied to his own behavior, the inconsistency would emerge. Years later (Campbell, 1987), he acknowledged that his commitment to evolutionary principles indeed seems to entail a pragmatist or contextualistic position, but insisted upon striving for the hypothetical realist goal of objectivity, even if objectivity can never truly be established. Campbell is not alone: evolution science in general has elevated objectivity above consistency with their own analytic concepts by failing to consider the scientist in the account.

Skinner faced the same conundrum but made a fundamentally different choice: to include an analysis of the contingencies controlling the scientist in his science (1945). This is what made his form of behaviorism "radical", undermining traditional behavioristic prohibitions against introspection, or an analysis of thoughts and feelings (for a discussion, see Hayes & Brownstein, 1986, and Friman, Hayes, & Wilson, 1998). This thoroughgoing naturalistic path leads us to a pragmatic departure from the traditional debates surrounding truth. We ask not how or whether genuine reference can possibly be established, but instead focus on what language and cognition do, rendering questions about correspondence superfluous to the account.

This approach is neither idealistic nor hostile to assumptions of a mind-independent reality. As Baldwin (1909) pointed out, acceptance of the principle of selection by consequences fits well with acknowledgement of an outside world. The inconsistency arises not from the ontology of mind-independence and representationalism, but from the realist's *epistemology* because selectivist accounts make unnecessary the work of providing guidance about *how one would know when* a truth-as-correspondence relation has been established. As Churchland put it, "natural selection does not care whether a brain has or tends toward true beliefs, so long as the organism reliably exhibits reproductively advantageous behavior" (1985, p. 45).

In our view, the assertion that a theory requires a special quality of reference in order to be instrumentally successful is not unlike the assertion that an organism needs a special vital force in order to live: it is an unnecessary intrusion of ontological assumptions into a functional approach (Roche & Barnes, 1997). To paraphrase Skinner (1969), scientific laws do not govern the universe; they govern the behavior of scientists coping with the universe. By transforming networks of stimulus functions, a theory can help us cope more effectively in and with the world. The additional claim that a theory is adaptive *because* it corresponds with the world adds nothing to the theory's utility. At its best, ontological speculation appears analogous to superstitious behavior— behavior not directly selected for, but which comes along for the ride accidentally. At its worst, ontology is a vestigial organ prone to infection—an appendix swollen with conceptual burden and controversy.

CBS and a Unified Contextualism

Scientific progress can be more vigorous when internally consistent perspectives are compared, because competition between ideas can proceed in a fashion that is more tightly linked to philosophy, principles, paradigms, and procedures and less linked to issues of personality, popularity, and persuasion. It is worth considering some of the features of CBS we listed earlier in light of our discussion of the basis of RFT in the evolution of a cooperative species.

- The core analytic assumptions of the philosophy of science underlying CBS are merely the assumptions implicit in variation and selective retention, as modified by the verbal purpose established by RFT. Truth is merely what works, but scientists are free to say, "Works toward what?" when considering their own verbal practices because that sets the criteria for selection. CBS as a community of scientists has done that, but in general evolution science has not laid out its goals in a similarly clear fashion. In the context of the legacy of social Darwinism, eugenics, and other horrors, it seems time for it to do so.

- RFT can likely benefit from evolution science and multi-level selection, but the reverse is also true. Genetic evolution occurs in the context of epigenetic, behavioral, and symbolic evolution (Jablonka & Lamb, 2005) and understanding more about the ontogenetic basis of cognition provides a focus and unit of analysis that has been missing from many evolutionist accounts of language. Phenotypic plasticity produced by learning and similar adjustments *drives genetic evolution itself* (West-Eberhard, 2003), which suggests that evolution science is incomplete without a selectivist account of language and cognition.

- The CBS idea of developing an applied theory using middle-level terms tied to basic principles makes perfect sense when one thinks of language pragmatically. Some forms of speaking are helpful in some contexts and other forms in other contexts. The use of a mix of clinically useful but looser terms (e.g., "acceptance") and technical terms that unpack these phenomena, translates this idea into CBS practices. Processes of

21

measurement development, component analysis, and mediation and moderation are empirical extensions of this idea.

- In the same way, early testing of effectiveness, dissemination, and training strategies is a clear extension of a selectivist perspective. Rather than assuming that understanding will lead to utility in a CBS approach, the precision and breadth of utility *is* understanding.

- Similarly, methods need to be tested across a broad range of areas and levels of analysis. Any failure of CBS to nest within other mature areas of science (e.g., evolution science) is a failure of CBS itself.

- Finally, the idea to create a cooperative development community based on these principles is an extension of multi-level selection processes to CBS itself. There should be no division between the content of the analysis and its application to those developing it.

Conclusion

Nesting CBS within evolution science makes better sense of its assumptions, methods, and strategy, and shows areas in which CBS may be of use to the larger community of selectivist accounts. We expect that the future will hold many elaborations of this basic vision and we hope that this brief exercise may be useful in that light.

References

Baldwin, J. M. (1909). *Darwin and the humanities*. Baltimore: Review Publishing Company.

Biglan, A., & Hayes, S. C. (1996). Should the behavioral sciences become more pragmatic? The case for functional contextualism in research on human behavior. *Applied and Preventive Psychology: Current Scientific Perspectives, 5,* 47-57.

Bloor, D. (1976). *Knowledge and social imagery*. London: Routledge & Kegan Paul.

Bond, F. W. Hayes, S. C., Baer, R. A., Carpenter, K. C., Guenole, N., Orcutt, H. K., Waltz, T. & Zettle, R. D. (2011). Preliminary psychometric properties of the Acceptance and Action Questionnaire—II: A revised measure of psychological inflexibility and experiential avoidance. *Behavior Therapy, 42,* 676–688.

Boyd, R. N. (1983). On the current status of the issue of scientific realism. *Erkenntnis, 19,* 45–90.

Breland, K., & Breland, M. (1961). The misbehavior of organisms. *American Psychologist, 16,* 681–684.

Campbell, D. T. (1959). Methodological suggestions from a comparative psychology of knowledge processes. *Inquiry, 2,* 152-83.

Campbell, D. T. (1960). Blind variation and selective retention in creative thought as in other knowledge processes. *Psychological Review, 67,* 380–400.

Campbell, D. T. (1987). Evolutionary epistemology. In G. Radnitzky & W. Bartley (Eds.) *Evolutionary epistemology, rationality, and the sociology of knowledge,* (pp.48-89). La Salle, Illinois: Open Court.

Cheney, D. L., & Seyfarth, R. M. (1990). Attending to behaviour versus attending to knowledge: Examining monkeys' attribution of mental states. *Animal Behaviour, 40,* 742-753.

Deacon, T. W. (1997). *The symbolic species: The co-evolution of language and the brain.* New York: Norton.

Churchland, P. M. (1985). The ontological status of observables: In praise of the supraempirical virtues. In P. M. Churchland & C. A. Hooker, (Eds.) *Images of science: Essays on realism and empiricism, with a reply from Bas C. van Fraassen.* Chicago: The University of Chicago Press.

Dougher, M. J., Hamilton, D., Fink, B., & Harrington, J. (2007). Transformation of the discriminative and eliciting functions of generalized relational stimuli. *Journal of the Experimental Analysis of Behavior, 88,* 179-197.

Foster, K. R., & Ratnieks, F. L. W. (2005). A new eusocial vertebrate? *Trends in Ecology and Evolution, 20,* 363–364.

Friman, P. C., Hayes, S. C., & Wilson, K. G. (1998). Why behavior analysts should study emotion: The example of anxiety. *Journal of Applied Behavior Analysis, 31,* 137-156.

Hacking, I. (1983). Experimentation and scientific realism. *Philosophical Topics, 13,* 71-87.

Hamilton, W. D. (1964). The genetical evolution of social behavior. *Journal of Theoretical Biology, 7,* 1–16.

Hare, B., & Tomasello, M. (2004). Chimpanzees are more skillful in competitive than in co-operative cognitive tasks. *Animal Behaviour, 68,* 571-581.

Hayes, S. C. (1992). Verbal relations, time, and suicide. In S. C. Hayes & L. J. Hayes (Eds.), *Understanding verbal relations* (pp. 109-118). Oakland, CA: New Harbinger/Context Press.

Hayes, S. C. (1993). Analytic goals and the varieties of scientific contextualism. In S. C. Hayes, L. J. Hayes, H. W. Reese, & T. R. Sarbin (Eds.), *Varieties of scientific contextualism* (pp. 11-27). Oakland, CA: New Harbinger/Context Press.

Hayes, S. C. (2008). Climbing our hills: A beginning conversation on the comparison of ACT and traditional CBT. *Clinical Psychology: Science and Practice, 15,* 286-295.

Hayes, S. C., Barnes-Holmes, D., & Roche, B. (2001). *Relational Frame Theory: A Post-Skinnerian account of human language and cognition.* New York: Plenum Press.

Hayes, S. C., & Brownstein, A. J. (1986). Mentalism, behavior-behavior relations and a behavior analytic view of the purposes of science. *The Behavior Analyst, 9,* 175-190.

Hayes, S. C., Hayes, L. J., & Reese, H. W. (1988). Finding the philosophical core: A review of Stephen C. Pepper's *World Hypotheses. Journal of the Experimental Analysis of Behavior, 50,* 97-111.

Hayes, S. C., Hayes, L. J., Reese, H. W., & Sarbin, T. R. (1993). (Eds.) *Varieties of scientific contextualism.* Oakland, CA: New Harbinger/Context Press.

Hayes, S. C., Levin, M., Plumb-Vilardaga, J., Villatte, J., & Pistorello, J. (in press). Acceptance and Commitment Therapy and contextual behavioral science: Examining the progress of a distinctive model of behavioral and cognitive therapy. *Behavior Therapy.*

Hayes, S. C., Luoma, J., Bond, F., Masuda, A., and Lillis, J. (2006). Acceptance and Commitment Therapy: Model, processes, and outcomes. *Behaviour Research and Therapy, 44,* 1-25.

Hayes, S. C., Strosahl, K., & Wilson, K. G. (2011). *Acceptance and Commitment Therapy: The process and practice of mindful change* (2nd edition). New York: Guilford Press.

Hölldobler, B., & Wilson, E. O. (1990). *The ants.* Cambridge, MA: Harvard University Press.

Jablonka, E., & Lamb, M. J. (2005). *Evolution in four dimensions: Genetic, epigenetic, behavioral, and symbolic variation in the history of life.* Cambridge, MA: MIT Press.

Kuhn, T. S. (1957). *The Copernican Revolution: Planetary astronomy in the development of Western thought.* Cambridge: Harvard University Press.

Kuhn, T. S. (1962). *The structure of scientific revolutions.* Chicago: The University of Chicago Press.

Laudan, L. (1977). *Progress and its problems: Towards a theory of scientific growth.* Berkeley: University of California Press.

Laudan, L. (1981). A confutation of convergent realism. *Philosophy of Science, 48*(1), 19-49.

Laudan, L. (1996). *Beyond positivism and relativism: Theory, method, and evidence.* Boulder, CO: Westview Press.

Levin, M., & Hayes, S. C. (2009). ACT, RFT, and contextual behavioral science. In J. T. Blackledge, J. Ciarrochi, & F. P. Deane (Eds.), *Acceptance and Commitment Therapy: Contemporary theory research and practice* (pp 1-40). Sydney: Australian Academic Press.

Levin, M. E., Hildebrandt, M. J., Lillis, J., & Hayes, S. C. (in press). The impact of treatment components suggested by the psychological flexibility model: A meta-analysis of laboratory-based component studies. *Behavior Therapy.*

Liebal, K., Behne, T., Carpenter, M., & Tomasello, M. (2009). Infants use shared experience to interpret a pointing gesture. *Developmental Science, 12,* 264-271.

McMullin, E. (1984). A case for scientific realism. In J. Leplin (ed.) *Scientific realism.* (pp. 8-40). Berkeley: University of California Press.

Moore, J. (2010). Philosophy of science, with special consideration given to behaviorism as the philosophy of the science of behavior. *The Psychological Record,* 60(1), 137-150.

Morris, E. K., Lazo, J. F. & Smith, N. G. (2004). Whether, when, and why Skinner published on biological participation in behavior. *The Behavior Analyst, 27,* 153-169.

Nowak, M. A. (2011). *Super cooperators: Altruism, evolution, and why we need each other to succeed.* New York: Free Press.

Nowak, M. A., Tarnita, C. E., & Wilson, E. O. (2010). The evolution of eusociality. *Nature, 466,* 1057-1062.

Pepper, S. C. (1942). *World hypotheses: A study in evidence.* Berkeley: University of California Press.

Putnam, H. (1978). What is realism? In H. Putnam, *Meaning and the moral sciences* (pp. 140-153). London: Routledge and Kegan Paul.

Radnitzky G., & Bartley, W. (Eds.) (1987). *Evolutionary epistemology, rationality, and the sociology of knowledge.* La Salle, Illinois: Open Court.

Roche, B., & Barnes, D. (1997). The behavior of organisms? *The Psychological Record, 47,* 597-618.

Roche, B., & Barnes-Holmes, D. (2003). Behavior analysis and social construction-ism: Some points of contact and departure. *The Behavior Analyst, 26,* 215-231.

Seyfarth, R. M., & Cheney, D. L. (2003). Signalers and receivers in animal communication. *Annual Review of Psychology, 54,* 145-173.

Skinner, B. F. (1945). The operational analysis of psychological terms. *Psychological Review, 52,* 270-276.

Skinner, B. F. (1969). *Contingencies of reinforcement: A theoretical analysis.* New York: Appleton-Century-Crofts.

Skinner, B. F. (1974). *About behaviorism.* New York: Knopf.

Skinner, B. F. (1977). Herrnstein and the evolution of behaviorism. *American Psychologist, 32,* 1006-1012.

Skinner, B. F. (1979). *The shaping of a behaviorist.* New York: Knopf.

Skinner, B. F. (1981). Selection by consequences. *Science, 213,* 501-504.

Skinner, B. F. (1984). Author's response. *Behavioral and Brain Sciences, 1,* 701-707.

Skinner, B. F. (1988). Genes and behavior. In G. Greenberg & E. Tolbach (Eds.), *Evolution of social behavior and integrative levels* (pp. 77-83). Hillsdale, NJ: Erlbaum.

Strosahl, K. D., Hayes, S. C., Bergan, J., & Romano, P. (1998). Assessing the field effectiveness of Acceptance and Commitment Therapy: An example of the manipulated training research method. *Behavior Therapy, 29,* 35-64.

Tomasello, M., Call, J., & Gluckman, A. (1997). The comprehension of novel communication signs by apes and human children. *Child Development, 68,* 1067-1081.

Tomasello, M. (2008). *Origins of human communication.* Cambridge, MA: The MIT Press.

Vilardaga, R., Hayes, S. C., Levin, M. E., & Muto, T. (2009). Creating a strategy for progress: A contextual behavioral science approach. *The Behavior Analyst, 32,* 105-133.

West-Eberhard, M. J. (2003). *Developmental plasticity and evolution.* New York: Oxford University Press.

Wilson, D. S. (2007). *Evolution for everyone.* New York: Bantam.

Wilson, D. S. (July, 2011). *Evolving the future: Toward a science of intentional behavior change.* Paper presented at the meeting of the Association for Contextual Behavioral Science, Parma, Italy.

CHAPTER TWO

The Pragmatic Truth Criterion and Values in Contextual Behavioral Science

Kelly G. Wilson

Kerry Whiteman

Michael Bordieri
University of Mississippi

Functional contextualism is a contemporary pragmatic philosophy that forms an organizing framework for a great deal of current behavior theory and empirical work. Truth occupies a strange space in contextual behavioral science (CBS), because it lies at odds with both commonsense notions of truth and, importantly, truth as it is understood in most mainstream science and philosophy. There are varieties of contextualism including both descriptive and functional contextualism (Hayes, 1993). Descriptive and functional contextualism both maintain that understanding an event can only occur in the context in which the event is embedded. However, descriptive contextualism differs in that its ultimate goals are not the prediction and influence of analyzed events, but instead an active personal appreciation of the functional relation between context and analyzed action. Examples of such efforts in psychology include dramaturgy, narrative psychology, and hermeneutics (Hayes, 1993). Our comments here will be restricted to functional contextualism because the focus of this volume is not consonant with a descriptive contextualist analysis.

In this chapter, we will argue that there is a received view of science that contains important epistemological and ontological assumptions related to truth that are not shared by functional contextualism. These assumptions in the received view of science, especially in psychology, are ubiquitous and largely unarticulated. Without understanding the contrasts between these positions, the notion of truth in CBS remains obscure. After defining the role of truth in CBS, we will explore applications of this criterion to practical settings as well as possible obstacles that might arise in the truth-seeking process. Lastly, we will conclude the chapter by outlining a number of suggestions that could help remediate the specified obstacles by applying these propositions to a real-world case study.

The Received View of Science

In order to properly understand "truth" in contextual behavioral science, we will first very briefly examine what is commonly meant by "truth" within mainstream psychological science. We will argue that there is a *received view of science* (RVS) that is 1) realist in its ontology, 2) convergentist in its epistemology, and 3) based on an assumed, but undefended and unarticulated referential theory of language. Furthermore, we will contend that these three features have a direct relation to misunderstandings of "truth" between CBS and much of RVS.

Hilary Putnam lays down the argument for RVS quite succinctly:

> The positive argument for realism is that it is the only philosophy that doesn't make the success of science a miracle. That terms in mature scientific theories typically refer ... that the theories accepted in a mature science are typically approximately true, that the same term can refer to the same thing even when it occurs in different theories—these statements are viewed by the scientific realist not as necessary truths but as part of the only scientific explanation of the success of science, and hence as part of any adequate scientific description of science and its relations to its objects. (Putnam, 1975, p. 73)

In what follows, we will compare and contrast each of these premises in light of the alternative worldview contained in CBS.

Worldviews: Reference, Realism, and Truth

Steven Pepper (1942) identified 4 relatively adequate worldviews, including formism, mechanism, organicism, and contextualism. Functional contextualism is an adaptation of Pepper's contextualist worldview developed by Hayes and colleagues in the 1980s (Hayes, Hayes, & Reese, 1988). A broad treatment of these worldviews is beyond the scope of this chapter, thus we will focus largely on the truth criteria of each, showing the critical contrast with truth as it is understood from a functional contextualist position.

In formism, the root metaphor is the commonsense activity of naming. As a scientific effort, the formist's task is to create a comprehensive set of categories. The adequacy of the analysis is to be found in the exhaustive nature of the categorical system. If the system has a category for all kinds of things, and things for all categories, then the categorical system is deemed to correspond with the *a priori* assumed world of things and events. It is important to note that the "thingness" of that which fills the categories is presumed to exist independent of the scientific endeavor.

Mechanism can be thought of as a more sophisticated variant of formism. Mechanism assumes the *a priori* status of parts, but goes on to build models involving parts, relations, and forces animating such a system. The truth criterion for this worldview is correspondence established through predictive verification. That is, the ability of the model to make remote predictions provides confirmation that the model adequately corresponds to the structural organization of the world that is analyzed. As with formism, parts, relations, and forces are assumed to exist independent of the scientific efforts to discover and model them.

Organicism, like formism and mechanism, identifies parts (i.e., stages of development), such as infant, toddler, child, adolescent, and adult. However, unlike formism and mechanism, organicism considers the whole to be primary and the parts "useful fictions" (see Hayes et al., 1988). That is, the parts or stages are seen as ways to organize scientific effort, but organicists recognize that the dividing lines are drawn arbitrarily for the purpose of investigation. The ordering of these somewhat arbitrarily delineated stages is, however, not at all arbitrary. In fact, the truth of an organicist theory lies in whether or not all data fit, or cohere, with the proposed developmental trajectory. Thus, where the line is drawn marking the difference between an infant and a toddler may be arbitrary, but that infancy precedes toddlerhood is nonarbitrary and is presumed to reflect the *a priori* organization of development. Again, as with previously described worldviews, the order of development is assumed to exist independent of the process of identifying that order and the theory is presumed to reflect *a priori* order.

Contextualism differs markedly. Truth in CBS is well captured by Skinner's dictum: science is "a corpus of rules for effective action, and there is a special sense in which it could be 'true' if it yields the most effective action possible" (1974, p. 235). Skinner places truth in scare quotes in order to emphasize his special use of the term. A theory is deemed true to the extent that it organizes the behavior of scientists such that it allows them to reach the goals of their science. In CBS, parts, relations, and forces may be described, but as with organicism, these parts are seen as a useful fiction. However, the described organization is not presumed to reflect the *a priori* organization of the world as in organicism.

Reference: Are Theories Mirrors or Hammers?

One way to consider the nature of theories metaphorically is to ask if theories are like mirrors or like hammers. It is largely assumed that the terms of theories mirror the world, albeit imperfectly. Consider Putnam's assertion that "terms in mature scientific theories typically refer" (1975, p. 73). In the full quotation above we see the implication that terms mirror reality when Putnam speaks of "adequate scientific description of science and its relations to its objects" (1975, p. 73). Presumably less mature theories mirror the *a priori* nature of reality less perfectly. Progress, by Putnam's definition, is equivalent to the polishing of our theoretical mirror so as to remove distortion and to more accurately re-present reality. Putnam contrasts this view with the idea that theory is merely an arbitrary narrative. How would we explain progress if there were not such a relationship between theory and reality? But, Putnam's argument only holds if one presumes that language is referential. Putnam's theory of theories rests on an unstated and unquestioned assumption that language itself genuinely refers to actual entities in the world.

Consider the alternative theory of theories. An alternate method of thinking of theories is to think of theories as instruments. A hammer is a "good" hammer if it allows the carpenter to drive a nail. It would not make sense to say that the hammer does so because it accurately refers to the nail or reflects the nail. This is the sense in which Skinner is speaking of theories. They are "true" if they allow the scientist to drive the nail, so to speak. In other words, theories are true to the extent that they help scientists achieve their goal. Of course, this is entirely consistent with Skinner's basic position on language itself. That is, Skinner rejected a referential theory of language and embraced a functional theory of verbal behavior (Skinner, 1957). Very early on, Skinner, in a very radical move, applied this functional analysis of language quite directly to the language of scientists (Skinner, 1945).

Realism

Skinner was interested in the contextual variables that allowed for the prediction and influence of behavior. Some aspects of the context are verbal, so it should be no surprise that his interest in verbal behavior is not whether or not it reflects reality. Instead, his interest focused on the ways in which verbal events are organized as behavior as well as how they can organize behavior. That is, Skinner contended that the meaning of language was to be found "among the determiners, not among the properties of response" (1945, p. 271). Skinner's functional approach to language included the verbal behavior we call theories and also, critical to this chapter, the verbal behavior that asserts the truth-value of a theory. A sensible Skinnerian analysis of the use of the word "true" would involve an analysis of the contexts that established and maintained scientists' use of the word "true".

Skinner's conclusion was that scientists use the term "true" when their theories lead to effective action and "not true" when and where they failed. He did not concern himself with the "ultimate" truth. He did not concern himself with the theory's accuracy in mirroring the world, because he had already rejected the copy-theory of ideas in a more general sense (Skinner, 1974) and its complement, a referential view of language (Skinner, 1945). He left metaphysics to philosophers and theologians. In fact, behavior theorists such as Willard Day have suggested that Skinner's position was "peculiarly anti-ontological" (Day, 1969, p. 319). In speaking of the behavioral scientist, Day states the following:

> In particular, he objects to speaking of the events associated in a functional relationship as if they were things and objects having a more or less permanent identity as real elements of nature. He does not believe that the functional relations he describes constitute an identification of anything which might be called "true laws of nature"... Rather, he is content for the most part simply to describe whatever natural consistency he can actually see, and to hope that the report he makes of his observations will in turn generate ultimately more productive behavior in the control of human affairs. (Day, 1969, p. 319)

Wilson (2001) also elaborates on this issue, making a distinction between operational and ontological validity. A theory might be said to have operational validity to the extent that it allows one to operate effectively on the subject matter of interest. The realist goes on to make what might be called an "ontological leap". That is, operational validity is assumed to be an indicator of ontological validity. This leap is central to predictive verification as described in the section above on mechanism. In a certain sense, the leap is necessary for the mechanist, and indeed for formism and organicism, since truth in all of these positions relies on accurate reference to antecedent conditions—that is, the "real" state of the world. The "true state" of the world is foundational for all worldviews apart from contextualism.

From a contextualist perspective, there are two problems with this ontological leap. The first is a logical error called *affirming the consequent* (If P then Q; Q; therefore P). The second and critical error is related to what pragmatists call the "cash value" of the added assertion. For the pragmatist, additional terms and assumptions are rejected unless they provide additional cash value. What is added by the ontological assertion that operational efficacy confirms good reference (i.e., that success confirms adequate mirroring of reality)?

For the realist, whose job is to uncover, or at least to converge upon, the "true nature" of the universe, it is a sensible leap. However, it is worth noting that this is a leap that has proven wrong with extraordinary regularity (see, Laudan, 1981, for strong refutation of RVS). For contextualists like Skinner, who are ready to leave metaphysics to the philosophers and theologians and be satisfied with solving problems, operational validity is all that is sought and all that is asserted. The ontological leap, which has no logical justification and no added "cash value", is therefore rejected.

Truth for the contextualist does not rely on the theory's more or less accurately mirroring antecedent conditions. The foundation for the contextualist is found not in the antecedents, but instead in the consequences. For the contextualist, greater and greater efficacy does not require any inference at all. Greater efficacy means greater efficacy and nothing more than that. This is quite a demotion for scientists. They are demoted from the role of discovering the most essential elements of the universe to being the solvers of problems. Sometimes those problems are of tremendous consequence, but still scientists' job remains that of a problem solver.

Scientists sometimes refer to the Chinese fable of the butcher whose knife never needed sharpening because he knew the anatomy of the animals so well that he always cut them at their joints. Some believe that the job of scientists is to cut nature at its joints. A good analysis may yield many good results and it may be tempting to assume that nature has indeed been cut at its joints. But consider the physics of light. In some experimental conditions, thinking of light as particles allows for optimal prediction. Under other conditions, thinking of light as waves makes better predictions (Bhuiyan, 2011). For the contextualist, no time need be spent concerning oneself with whether or not light *is* a wave or a particle. One adopts the most fruitful language given the context. Of course, striving for precision, scope, and parsimony, the contextualist would value the development of a single way of speaking that was effective in both experimental conditions. However, even if successful, she would not conclude that this new language *mirrored* reality or *referred* to the "true" state of the world. The contextualist holds persistently to the assumption that the true analysis can only be known in the solving of problems.

Functional Contextualism, Values, and the Two-Pronged Problem of Subjectivity and Relativism

Two potential problems emerge with functional contextualism. The first is the problem of subjectivity, or the fact that a scientist's perspective can heavily influence his or her analysis and conclusions. The second is the problem of relativism, which refers to changeable "truth" based on a given scientist's personal goals (cf. Barnes-Holmes, 2000). Both of these issues arise from the very nature of the truth criteria of functional contextualism. Hayes' attempt to make contextualism a more coherent and useful scientific strategy involved the rejection of what he called dogmatic goals for contextualists (Hayes, 1993). He suggested that goals of a science, which are necessary to mount the truth criterion of effective action, cannot themselves be defended except by appeals to other goals. For example, if I said my goal was to end disease, I could be asked why that was important. If I argued that it preserved life, I could be challenged on why that was important. If I argued that it was based on survival of the human species, someone might ask for a justification for that goal, *ad infinitum*. The number would be few, but some might argue that humans are a blight on the planet and that their continued existence is a pestilence. Hayes argued forcefully that functional contextualists need to simply state their goals without justification and allow the evaluators of the scientific effort to accept the goal and evaluate effective action according to those goals, or to "vote with their feet" (Hayes, 1993, p. 18)—that is, they could disagree with your goals and denounce or simply ignore your scientific efforts.

Since goals in functional contextualism are not accepted dogmatically, they must, per force, be selected on a subjective basis. One might accept them based upon group consensus, but accepting group consensus implies that one values the group's views, which must, by Hayes' argument, be an undefended choice.

This issue of values is no small matter, since the whole of "truth" in functional contextualism rests upon it. Therefore, it is worthwhile to attempt to unpack the term "values" from a functional contextualist perspective. The most recent functional analysis of values from an explicitly functional contextualist view can be found in Wilson and DuFrene (2009; cf. Hayes, Strosahl, & Wilson, 2012):

> *Values* are freely chosen, verbally constructed consequences of ongoing, dynamic, evolving patterns of activity, which establish pre-dominant reinforcers for that activity that are intrinsic in engagement in the valued behavioral pattern itself. (Wilson & DuFrene, 2009, p. 66)

Freely chosen, in this context, does not mean undetermined. It is meant in the sense that Skinner meant free, which is behavior not under aversive control (Skinner, 1971). Verbally constructed is intended to suggest that values are a type of rule about a pattern of activity. Establishing reinforcers is intended to suggest the type of rule—that is, an augmental, or rule that creates or increases the reinforcing potency of some reinforcer. Values furthermore entails the suggestion that the predominant reinforcer is intrinsic in the pattern of action. For example, good parenting might be a value an individual has declared. An array of potential outcomes might be identified, such as good social, health, and economic outcomes. However, parents of children with profound disabilities or fatal diseases often continue to value parenting even though those outcomes are unavailable. Simple engagement in the act of parenting is enough. Two other features of this definition are worth noting as we scale from the level of a therapeutic principle to an analysis of organized scientific efforts. Namely, a value is a "dynamic, evolving" pattern of activity. That is, the definition recognizes the recursive nature of valued action from this perspective. In addition, valuing as a behavioral act itself is seen as historically situated and functionally related to an "ongoing" stream of interaction with environmental conditions. In this sense, values are not static in that they can change with the addition of new, relevant experiences.

The Problem of Subjectivity

Personal values have often been seen as the enemy of good science, inasmuch as they are viewed as sources of bias. Seventy years ago, some, such as Stevens (1939), pursued the notion that we might have a psychological science that was purely objective and presumably free of the biases inherent in values. A science made up of only empirical and logical sentences would be capable of "blasting a path through subjectivity" (Stevens, 1939, p. 231). That vision, though grand enough, is fraught with problems, both practical and logical.

Feminist sociologist Jessie Bernard (1987) addresses this issue quite clearly in the prolegomenon of her book *The Female World from a Global Perspective*:

> This book has not turned out to be the book I originally envisaged, a book my kindly disposed but critical friends called "visionary." The "vision" I had in mind was analogous to a Mercator projection of the female world without the distortions such a projection of a sphere imposes on a flat plane. It was, in brief, to be a view of the female world from a perch that could offer a global view unbiased by its location, an admittedly "mission impossible." Among the countless obstacles in the way of such a "vision" were some that were semantic in nature and some intrinsically biasing. (p. vii)
>
> I am well aware that I am not a blank page onto which data fall, self-organized and already interpreted, ready for my typewriter: They come, rather, from a computer programmed by researchers who, like me, asked it certain questions and fed it certain variables. The data may therefore never have been touched by human hands, but human minds were guiding the process from start to finish, and these minds were no more virginal than my own. (p. x)

Bernard calls our attention to the inescapable fact that observation sentences are themselves theory laden. And that our theories *come from somewhere*—from a particular sociocultural context. Bernard's view, though perhaps not a functional contextualist view, is certainly a contextualist view. If one takes Bernard seriously, the best we can do is to attempt to articulate our biases, that is, the context from which we speak, while recognizing that even our biases are viewed from a perspective. From a functional contextualist view, a critical element of the context is the purpose of the analysis. That is, a scientist's goals are a critical element of the sociocultural context that influences scientific investigation, and the products of scientific inquiry cannot be properly evaluated without knowledge of the scientist's position and purpose.

To the extent that these contextual variables are public matters, we are in a much better position to know what biases to expect, and potentially to guard against (see Ruiz & Roche, 2007). It is worth mentioning that Bernard's as well as Ruiz and Roche's critiques are not restricted to contextualists. Typically, the place of values in science has been obscured by the myth that science has none (i.e., science for science's sake). While even if it were true that the science has none, it does not follow that the scientists do not.

Relativism, Subjective Bias, and Functional Contextualism

The other major potential concern with contextualism is that of relativism (Ruiz & Roche, 2007). When truth is defined by effective action, one must always define truth against action to particular ends. Problems with the motives of scientists are hardly restricted to functional contextualism, since the history of RVS contains a number of instances of Machiavellian science. The eugenics movement provides an apt example. Thus, the problem of relativism has not historically been prevented by adherence to realism or to a referential theory of theories.

The functional contextualist is still concerned about bias—not that it exists at all, but about biases that prevent approaching goals. For example, if one's goal is to alleviate starvation, being biased in favor of adequate nutrition is not a problem. However, it is not difficult to imagine biases that might undermine this goal. For example, the scientist might launch a vitriolic public campaign that actually caused more people to turn away from the cause, resulting in *increases* in hunger. The content of the rhetoric might be correct. However, exercise of these biases in this way might be corrected in such a way that more gains could be made.

In another scenario, imagine the scientist develops an effective intervention strategy. The success of the strategy might well advance the likeliness that the scientist would publish in high-level journals, find recognition among peers, citations in text books, and financial incentives in the form of a higher salary, for example. These reinforcers are not intrinsically problematic. After all, the scientist may have just as much at stake in putting food on her family's table as she does in solving the global hunger crisis. However, it's fairly simple to see how external contingencies such as money and fame might be in conflict with the goals of the science.

If one takes seriously that even benign sounding empirical statement is theory and value laden, it becomes clear that mechanisms must be put in place to address these biases. Science as a more general enterprise has methods to mitigate against these biases. Science is somewhat adversarial. If one produces a theory of any note, opposition will arise. Someone else will have his or her own theory and will begin gathering evidence to refute the original theory. Here we see potential bias on the part of both scientists, but the biases may help the scientific effort by setting up fruitful comparison of accounts. Within scientific disciplines, that adversarial relationship is facilitated by certain practices. For example, high specification of methods ensures that others can replicate a researcher's findings. Or, another researcher may see a confounding variable in those methods that was not apparent to the original theorist. At times, researchers find strong support for an opposing theory. These unexpected findings are themselves compelling evidence for science as a self-correcting enterprise. Other practices, such as peer review, work to lessen the chance that science will be blinded in important ways by the perspective of the scientist.

There are, however, ways in which this process can fail. For example, if views are extremely pervasive and that pervasiveness is enshrined in decision-making processes at funding agencies, academic departments, and scientific journals, it may be very hard for alternate views to emerge. The area of genetics provides a good example. There was a period of time when none other than Nobel laureate Francis Crick called a segment of the human genome junk DNA:

> The conviction has been growing that much of this extra DNA
> is "junk," in other words, that it has little specificity and conveys

little or no selective advantage to the organism (Orgel & Crick, 1980, p. 604).

We now know that these hitherto *junk* areas play a crucial role in the workings of our genes. The following exchange between geneticist Phillip Zamore and reporter Joe Palca on National Public Radio illustrates the problem:

Mr. ZAMORE: And there were whole classes of genes we had missed for the 40 years of intense molecular investigation. And people go around at meetings saying to each other, "Remind me again how we managed to miss this?"

PALCA: So how did they manage to miss it?

Mr. ZAMORE: I—you know, science is like every other endeavor: You see what you're trained to see. (Palca, 2005)

This tendency to seek out evidence that supports our own expectations while ignoring or downplaying contradicting evidence is known as the confirmation bias and is a ubiquitous phenomenon in scientific inquiry (Nickerson, 1998). These observations in genetics almost certainly impacted research careers, funding streams, and the publishability of findings. When we see such things in an area like genetics, it is a small leap to imagine how factors like gender, ethnicity, and culture might play a role in what kinds of questions are asked, what answers are dismissed or listened to, whose research program is funded and whose is marginalized.

From a contextualist perspective, biases are assumed to exist, for the precise reasons Jesse Bernard outlines above. There is no seeing that does not occur from a particular current and historical perspective. All seeing is situated seeing and, just as Skinner emphasized in 1945, the scientist has no privileged perch. The question is, can we put in place procedures that make the revelation of biases subject to correction sooner rather than later? The difference between functional contextualism and the received view of science is that for most of science, bias is a nuisance variable. For the contextualist, it is foundational inasmuch as meeting the specific goals of analysis is foundational to truth.

Truth Finding, Functional Contextual Science, and the Received View of Science

In any science, the matter of "truth" finding is centrally important. Most ordinary science practices apply readily to the doing of contextual science. However, the interpretation of the results of typical scientific practices might be interpreted differently from a contextualist view. For example, independent replication of a finding tells the mechanistic scientist about the ontological validity of the model. The functional contextualist will make a more circumspect assertion and might conclude that *so far* as the analysis extends to that other setting and group of scientists, the description of methods was adequate to reproduce the finding. However, the reproduction of that finding might also be interpreted as reflecting shared biases that restrict the "truth value" of the theory. For example, other cultural contexts might reveal different results. Those different results might in turn alter theory in such a way as to extend to both cultural contexts. From the CBS perspective, the discovery of useful scientific principles is enhanced by diversity of inquiry and cross-cultural collaboration early on in the process, as such collaboration aids in the identification of cultural biases that limit the utility of scientific findings.

Solving the Potential Problems with Truth in Functional Contextualism: A Case Study with Critique and Suggestion

Recently, Hayes and colleagues (in press) have made some suggestions for ways that contextualism could circumvent these particular difficulties and progress as an explicit basis for a scientific enterprise:

1. Avoid implicit goals.

2. Avoid vague goals.

3. Avoid incompatible goals.

4. Avoid using solely short-term goals.

5. Avoid extremely long-term goals, without medium- and short-term goals.

6. Avoid rapidly changing goals.

7. Compare performance to goals.

8. Compare different courses of action and modify behavior accordingly.

Rather than consider these suggestions in the abstract, we will examine contextualism in practice. In what follows, we briefly examine the only substantial functioning organization that explicitly espouses functional contextualism as a guiding philosophy: the Association for Contextual Behavioral Science (ACBS).

ACBS was founded in 2005 by behavioral scientists and practitioners interested in behavior analysis, functional contextualism, relational frame theory, and acceptance and commitment therapy. The society has grown to over 4,000 members worldwide.

The most substantial sources of information on the society can be found in the public domain on their website (www.contextualpsychology .org) and information in the following sections are based on that information.

The first of Hayes and colleagues' suggestions is that contextualists should avoid implicit goals. Explicit goals avoid potential problems posed by Ruiz and Roche (2007). To the extent that the goals are public, they are subject to broad evaluation. Below are the stated goals of ACBS:

ACBS is dedicated to the advancement of functional contextual cognitive and behavioral science and practice so as to alleviate human suffering and advance human wellbeing. This organization seeks:

• The development of a coherent and progressive science of human action that is more adequate to the challenges of the human condition.

- The development of useful basic principles, workable applied theories linked to these principles, effective applied technologies based on these theories, and successful means of training and disseminating these developments, guided by the best available scientific evidence.

- The development of a view of science that values a dynamic, ongoing interaction between its basic and applied elements, and between practical application and empirical knowledge.

- Development of a community of scholars, researchers, educators, and practitioners who will work in a collegial, open, self-critical, non-discriminatory, and mutually supportive way that is effective in producing valued outcomes and in exploring the additional implications of this work, and that emphasizes open and low cost methods of connecting with this work so as to keep the focus on benefit to others. (Association for Contextual Behavior Science, 2009)

These explicit and publicly available goals serve three critical functions. First, if potential supporters of the science do not agree with the goals, they can oppose the work. Second, to the extent that contextualists' goals are made explicit, they are subject to examination by a larger and more diverse audience. Human judgment is demonstrably sensitive to biasing effects. Experimental psychology, and social psychology in particular, is rife with examples of the ways in which human judgment can be biased. Primacy, recency, prior allegiance, cultural factors, and phrasing of statements have all been demonstrated to impact human judgment (Choplin, 2010; Fowler, 1996; Hergovich, Schott & Burger, 2010; Kerr, Ward & Avons, 1998). Even agreement by a large number of individuals is no guarantee that biases will be avoided (Choi & Choi, 2010). To the extent that the persons agreeing come from a similar point of view, they are likely to share biases. The importance of diverse perspectives is not because others will not have biases, but because, to the extent that they have different histories, they are less likely to have *the same biases*.

The third critical function explicit and public goals yield can be examined in regard to some of the other suggestions that Hayes (1993) describes. More specifically, to the extent that the goals are public, they can be examined for desirable levels of both flexibility and stability. The

meeting of various long- and short-term goals, including their explication, can be met by the availability of evidence. Current evidence can
primarily be accessed through peer reviewed scientific journals.

Improving the Pursuit of Truth from a Functional Contextual Perspective

What follow are a series of practical suggestions that would apply equally
to any functional contextual science effort. These suggestions are applied
specifically to ACBS as it is currently the only explicitly contextualist
scientific society. This makes it a worthy test bed for the evolution and
application of contextual science principles.

Increase transparency of membership. One key element in evaluating
the goals of a group is in knowing the constituency. For example, if we
see a group promoting knowledge of tobacco use and health risks, we
might be more skeptical if we found out that all its members were individuals from the tobacco industry. Contextualist organizations like
ACBS should make efforts to improve the transparency of membership.
Although members are accessible through the membership directory, it is
not in a form that readily allows an observer to get a sense of *who* ACBS
is. A quarterly snapshot of ACBS membership would help in the evaluation of the society and its goals. Such a snapshot might include information about 1) country of origin, 2) language groups represented, 3) ethnicity, 4) type of profession or public membership, and 5) gender, for
example. These demographics could help nonmembers better evaluate
ACBS. It would also allow ACBS to better target outreach efforts to the
broadest possible constituency.

Make funding transparent. An additional related issue is the source of
funding for ACBS. When we know where the money comes from, we are
often better able to understand the message. For example, an individual
giving a talk on the benefits of a pharmaceutical agent might be viewed
differently if the audience knew that he or she was being paid by the
pharmaceutical company that manufactured the drug. Although ACBS
is currently funded only by membership dues, opening the books of the

society and making funding publicly and readily available would aid in transparency.

Increase breadth of constituency. ACBS was established as a collaboration among clinicians and scientists. This sort of effort has the potential to bridge the science-practice divide in psychology. Along these lines, recent connections to evolutionary science are suggestive of broadening coherent cross-disciplinary perspectives. Chapters among different language groups and regions are beginning to emerge. Increased transparency of membership, as described above, could help the organization better target under-represented groups and regions.

Although ACBS is open to membership by the public, the overwhelming majority of members are mental health professionals and researchers. Contextual science efforts of "truth seeking" aimed at the improvement of the human condition would be well-served by greater public involvement. A potential function of greater outreach would be the dissemination of evidence-based principles among a public that is often informed by corporate interests focused on profit-making. Another potential outcome might be the broadening of the communitarian effort into the pool of the most important constituents: the consumer of behavioral health applications.

Increase accessibility of goals and progress. ACBS goals are currently available to both public and professional constituencies. The progress of contextual behavioral science efforts is less accessible. Progress is now available in the form of peer reviewed articles and chapters in scholarly texts. Because of the cost of access and the understandability of technical writing, many of these are of limited use to nonprofessionals. In order to have the maximum scrutiny and input, the evidence base needs to be translated into a form that is more accessible to individuals outside academic audiences.

Instantiate in bylaws a regular recursive review of the goals and progress. Hayes suggests a recursive process in which the achievement of short- and long-term goals, and even the goals themselves, is subject to recurrent evaluation and potentially to revision (Hayes, 1993). Although Hayes warns about too rapidly changing goals, a systematic and established pathway to revision seems appropriate for an organization such as ACBS. The scrutiny and refinement of goals could be best facilitated by the broadest possible stakeholder participation. Likewise, the systematic occurrence of such a town hall meeting would avoid complacency on the part of ACBS leadership, as the organization becomes more established and potentially more resistant to change.

Conclusion

Contextual science is in some regards a new effort in experimental science. Although the underlying functional model is over a century old, the mechanistic received view of science has been far more dominant. The examination of and suggestions for contextual science efforts may appear out of place in a chapter aimed at describing the nature of truth from a contextualist view. The pragmatic truth criterion found in contextualism lends itself to any sort of goals, and thus any reference point for truth. Therefore, a broad communitarian effort is especially important in ensuring the ongoing integrity and accountability of contextual science.

Science can be thought of as a public trust. The doing of science is often funded by the largess of taxpayers. Taxpayers likewise often subsidize the training of scientists. Taking this view, the general public has an investment in the products of science. There is perhaps no more critical science than the science of human behavior. It is simply too important to be left solely in the hands of scientists or as the property of scientists. *The science of human behavior is everybody's business and nobody's property.* In a very important sense, we are all stakeholders in the science of human behavior. Contextual science, though it opens the door to relativism, also provides a philosophy of science rationale for a communitarian approach to the monitoring and development of that science.

References

Association for Contextual Behavior Science. (2009, April 28). About us. Retrieved from http://contextualpsychology.org/acbs

Barnes-Holmes, D. (2000). Behavioral pragmatism: No place for reality and truth. *The Behavior Analyst, 23,* 191-202.

Bernard, J. (1987). *The female world from a global perspective*. Indiana: Indiana University Press.

Bhuiyan, A. L. (2011). The genesis of wave-particle duality. *Physics Essay, 24*, 16-19.

Choi, D. & Choi, I. (2010). A comparison of hindsight bias in groups and individuals: The moderating role of plausibility. *Journal of Applied Social Psychology, 40(2)*, 325-343.

Choplin, J. M. (2010). I am "fatter" than she is: Language-expressible body-size comparisons bias judgments of body size. *Journal of Language and Social Psychology, 29(1)*, 55-74.

Day, W. F. (1969). Radical behaviorism in reconciliation with phenomenology. *Journal of the Experimental Analysis of Behavior, 12(2)*, 315-328.

Fowler, R. D. (1996). 1996 editorial: 50th anniversary issue of the American Psychologist, *51*, 5-7.

Hayes, S. C. (1993). Analytic goals and the varieties of scientific contextualism. In S. C. Hayes, L. J. Hayes, H. W. Reese & T. R. Sarbin (Eds.), *Varieties of scientific contextualism* (pp. 11-27). Reno, NV: Context Press.

Hayes, S. C., Hayes, L. J. & Reese, H. W. (1988). Finding the philosophical core: A review of Stephen C. Pepper's World hypotheses: A study in evidence. *Journal of the Experimental Analysis of Behavior, 50(1)*, 97-111.

Hayes, S. C., Levin, M., Plumb, J., Villatte, J., & Pistorello, J. (in press). Acceptance and Commitment Therapy and contextual behavioral science: Examining the progress of a distinctive model of behavioral and cognitive therapy. *Behavior Therapy*.

Hayes, S. C., Strosahl, K., & Wilson, K. G. (2012). *Acceptance and commitment therapy: The process and practice of mindful change* (2nd ed.). New York: Guilford Press.

Hergovich, A., Schott, R., & Burger, C. (2010). Biased evaluation of abstracts depending on topic and conclusion: Further evidence of a confirmation bias within scientific psychology. *Current Psychology, 29*, 188-209.

Kerr, J., Ward, G., & Avons, S. E. (1998). Response bias in visual serial order memory. *Journal of Experimental Psychology: Learning, Memory, and Cognition, 24(5)*, 1316-1323.

Laudan, L. (1981). A confutation of convergent realism. *Philosophy of Science, 48(1)*, 19-49.

Nickerson, R. S. (1998). Confirmation bias: A ubiquitous phenomenon in many guises. *Review of General Psychology, 2*, 175-220.

Orgel, L. E., & Crick, F. H. (1980). Selfish DNA: The ultimate parasite. *Nature, 284(5757)*, 604-617.

Palca, J. (2005, October 10). DNA: The machinery behind human beings—Part 1. *National Public Radio*. Retrieved from http://www.npr.org/templates/story / story.php?storyId=4953053

Pepper, S. C. (1942). *World hypotheses: A study in evidence*. Berkeley: University of California Press.

Putnam, H. (1975). *Matter, Mathematics and method: Philosophical papers, volume 1.* Cambridge University Press: Cambridge.

Ruiz, M. R., & Roche, B. (2007). Values and the scientific culture of behavior analysis. *Behavior Analyst, 30*(1), 1-16.

Skinner, B. F. (1945). The operational analysis of psychological terms. *Psychological Review, 52,* 270-277.

Skinner, B. F. (1957). *Verbal behavior.* New York: Appleton-Century-Crofts.

Skinner, B. F. (1971). *Beyond freedom and dignity.* New York: Alfred A. Knopf.

Skinner, B. F. (1974). *About behaviorism.* New York: Appleton-Century.

Stevens, S. S. (1939). Psychology and the science of science. *Psychological Bulletin, 36,* 221-263.

Wilson, K. G. (2001). Some notes on theoretical constructs: Types and validation from a contextual-behavioral perspective. *International Journal of Psychology and Psychological Therapy, 1,* 205-215.

Wilson, K. G., & DuFrene, T. (2009). *Mindfulness for two: An acceptance and commitment therapy approach to mindfulness in psychotherapy.* Oakland, CA: New Harbinger.

PART II

Advances in Basic Research on Relational Frame Theory

CHAPTER THREE

Relational Frame Theory:
An Overview

Ian Stewart

National University of Ireland, Galway

Bryan Roche

National University of Ireland, Maynooth

Introduction

Understanding language is crucially important to understanding human psychology. For this reason, psychologists need an account of language that is scientifically adequate. From the current perspective, such an account is one that is based on empirical research and provides a bottom up explanation of relevant processes of interaction between behavior and environment, starting with the simple and advancing incrementally to the complex, thereby producing a terminology and accompanying methodologies that allow both prediction and influence with respect to that phenomenon. We believe that Relational Frame Theory (RFT) constitutes such an account. The purpose of this chapter is to provide a brief overview of RFT, starting with its empirical background and then explaining its basic tenets.

Background to Relational Frame Theory

RFT is grounded in behavior analysis, an empirically well rooted and practically oriented approach to psychology and one that has achieved substantial success in certain applied arenas (e.g., early intensive behavioral intervention, organizational behavior management). Despite the latter, behavior analysis is not currently a mainstream approach to psychology. One key reason for this is that for a long time in its history it has had a reputation for failing to offer a potentially adequate account of complex human behavior and in particular language.

The major behavior analytic conceptualization of language before RFT was provided over half a century ago by Skinner (1957), who attempted to approach it by extrapolating from direct operant responding, which had been researched in non-humans. The core of Skinner's thesis was that verbal behavior was reinforced through the mediation of a listener specially trained to provide such reinforcement. However, there are fundamental problems with this definition. First, it is non-functional, since the analysis of the behavior of the speaker relies on knowledge of the history of another organism, the listener. Second, it is too broad. In an operant experiment, the behavior of the organism under investigation is reinforced by an experimenter specially trained to do so; thus, by Skinner's definition the behavior of non-humans in such experiments is verbal. Third, while the behavior of the speaker is considered verbal, the behavior of the listener is not, which seems to at least downplay the importance of verbal comprehension as opposed to production.

Apart from the problems posed by these theoretical issues in themselves, at least one of them also led directly to serious problems in terms of empirical analysis. Since organisms in operant experiments were already engaging in verbal behavior, then it proved impossible to isolate the latter in order to study it. Hence Skinner's analysis failed to produce a comprehensive and progressive program of research (see e.g., Dymond & Alonso-Álvarez, 2010). This was a fundamental problem and is a key reason that the Skinnerian account has proven inadequate.

As it turns out, if Skinner had not extrapolated from non-human operant responding as early as he did, but had waited for the arrival of additional data on human responding then a very different account of human verbal behavior might have resulted, because in the two decades following Skinner (1957), data from two areas of research in particular indicated possibly substantial differences between humans and other organisms that appeared relevant to the understanding of language.

Differences between Human and Non-Human Behavior

The first of these areas was performance on intermittent schedules of reinforcement. In intermittent schedules, reinforcement is not provided for every response but only for patterns of responding that conform with predefined parameters (e.g., high rate). It has typically been found that the responding of the organism under investigation (in traditional experimental research typically a rat or pigeon) gradually adjusts to the contingencies of reinforcement so that eventually its behavior is relatively effective with respect to the particular schedule. However, research on intermittent schedules with verbal humans (typically with points exchangeable for money as reinforcement) shows significant and consistent differences between their behavior and that of non-humans (e.g., Galizio, 1979). Furthermore, the evidence suggests that the former respond differently because they are covertly following rules about how to respond. For example, in order to investigate the possible effect of rules on schedule responding, researchers have deliberately provided human participants in schedule experiments with rules about the best way to respond and have shown unambiguous and consistent effects (see Hayes, 1989, for an overview).

Research on intermittent schedules of reinforcement has thus provided one clear empirical demonstration of a substantial difference between verbal humans on the one hand and non-verbal subjects on the other: the first group show "rule-governed behavior" which can hugely affect schedule performance. Empirical evidence of the influence of self-created or experimenter-provided rules on human schedule performance first began to emerge in the 1960s, and investigation of this phenomenon continued throughout the next few decades. Despite this interest, however, throughout this time, no adequate technical understanding of what rules were or how humans understood or followed them was developed. Skinner (1969) described rules as "contingency specifying stimuli", but failed to provide a behavioral definition of "specify" (O'Hora & Barnes-Holmes, 2001). An adequate understanding of verbal behavior might have seemed relevant. However, there was no behavior analytic account of language that could help. Skinner's account of verbal behavior was the only one in existence when rule-governed behavior began to be researched. However, explanation of the latter requires understanding verbal comprehension; in other words, how the rule affects the listener, and, as explained previously, Skinner's definition of verbal behavior excluded the behavior of the listener and hence could shed no light in this regard.

A second area of research, investigating conditional discriminative responding, revealed another substantial difference between verbal humans and other groups. In typical conditional discrimination training, the experimental subject is taught to choose one of a number of discriminative (comparison) stimuli contingent on the presence of one of a number of conditional discriminative (sample) stimuli (see, e.g., Saunders and Williams, 1998). The study which first provided evidence of a substantial behavioral difference between verbal humans and other groups in this context was Sidman (1971). The original aim of this study was to improve the reading comprehension of a developmentally disabled individual by training a series of related conditional discriminations involving spoken words, pictures and printed words. The individual could already select appropriate pictures (B) in the presence of particular spoken words (A) and could utter the correct names (A) in the presence of the pictures (B), and at the start of the study he was taught to select the appropriate printed words (C) in the presence of the spoken words (A). However, in subsequent testing he demonstrated not just these already established performances but several additional untrained or derived performances that conventional behavior analytic theory would have not have predicted including utterance of the correct names (A) in the presence of the appropriate printed words (C), selection of the pictures (B) in the presence of the printed words (C) and of the printed words (C) in the presence of the pictures (B). Sidman suggested that this outcome indicated that he was now treating spoken words, pictures and printed words as mutually substitutable or equivalent, and thus he termed this phenomenon "stimulus equivalence."

Sidman argued, on the basis of these and similar findings, that stimulus equivalence was defined by three properties: *reflexivity*, or identity matching (i.e., selecting a stimulus A in the presence of another identical stimulus A); *symmetry*, or reversal of the conditional discriminative functions of sample and comparison (e.g., in the above, the subject was taught to select printed words (C) in the presence of spoken words (A) and subsequently produced the correct names (A) in the presence of printed words (C)); and *transitivity*, or derived conditional discriminative responding based on the training of two or more other conditional discriminations (e.g., in the above, the subject showed the derived performances of selection of pictures (B) in the presence of printed words (C) and vice versa, based on the two previously established repertoires of selection of pictures (B) in the presence of spoken words (A) and of printed words (C) in the presence of spoken words (A)).

The equivalence phenomenon has garnered much research attention since Sidman (1971) for several reasons. For one thing, as suggested, it was not predicted by conventional theory; thus, additional research was directed to discovering more about its parameters and the circumstances under which it appeared. In addition, it appeared to be of great practical benefit. Sidman had been using it to train reading comprehension and had demonstrated that only a subset of the conditional discriminative performances required for such comprehension needed to be directly trained in order for the full set to emerge.

Perhaps the most compelling reason for basic researchers to examine the phenomenon of stimulus equivalence, however, was its apparent connection with language. The stimulus substitutability and generativity characterizing equivalence resemble those of language and suggested that this phenomenon would provide a useful model of the latter. An accompanying empirical effect referred to as transfer of function, whereby a psychological function trained up in one stimulus in a derived equivalence relation with another subsequently appears in the latter without training (e.g., Dougher, Augustson, Markham, Greenway, & Wulfert, 1994), further bolstered this suggestion as it appeared to offer an analog of linguistic control. In addition, a range of other empirical findings since Sidman's original study have provided further evidence that stimulus equivalence and language are closely linked (e.g., Barnes, McCullagh, & Keenan, 1990; Ogawa, Yamazaki, Ueno, Cheng, & Iriki, 2009).

A variety of theories have emerged from within behavior analysis to explain the link between language and equivalence. However, from the current perspective, the explanation provided by RFT is the most scientifically adequate. Furthermore, RFT also provides an adequate behavioral explanation of rules. In this way, it can account for both the empirical effects suggesting a difference between verbal humans and non-verbal organisms that have just been discussed. It does this by explaining the nature of the behavioral phenomenon that underlies human language, arguing that it is the same as that which underlies stimulus equivalence. Furthermore, there is by now a substantial level of empirical support for this conception, provided within the context of an expanding program of basic and applied research into verbal processes from the RFT point of view. As such, though the RFT account remains the subject of some continuing debate within behavior analysis itself (see Gross and Fox, 2009), we believe that it constitutes a critically important theoretical and empirical advance for behavior analysis and for scientific psychology in general. In what follows, we will explain the RFT approach to language and related phenomena including derived relations, rules and complex behavior more generally.

Theoretical Foundations of Relational Frame Theory

Relational Frame Theory explains the link between language and stimulus equivalence by suggesting that they are the same phenomenon, namely, arbitrarily applicable relational responding. In order to explain this phenomenon we need to first consider relational responding.

Non-Arbitrary Relational Responding

In relational responding, a subject shows a generalized repertoire of responding to one stimulus in terms of another stimulus (see e.g., Hayes, Fox, Gifford, Wilson, Barnes-Holmes & Healy, 2001). A classic example of this is identity matching. Identity matching can be demonstrated using a conditional discriminative procedure in which the correct comparison is physically identical to the sample. For example, the conditional discriminative (sample) stimuli in a matching-to-sample task might be a red circle and a green circle and the potential discriminative (comparison) stimuli might be a red circle and a green circle. On a trial in which the red circle appears as the sample, the correct comparison is the red circle; on a trial in which the green circle appears as the sample, the correct comparison is the green circle. However, it is not enough for the purposes of demonstrating identity matching that this performance be displayed on the basis of training with specific sets of stimuli. For true identity matching to be displayed, the subject must show generalization of the performance to novel sets of stimuli such that he or she consistently chooses on the basis of a relation of physical (in this case color) similarity between the sample and comparison.

Identity matching is also referred to as non-arbitrary relational responding (NARR; also called "transposition"; see Reese, 1968) because the subject responds based on a relation between non-arbitrary or formal properties of the stimuli being related. Such properties (e.g., color, shape, quantity, size) are referred to as non-arbitrary because they are not subject to social whim, as contrasted with arbitrary or arbitrarily applicable properties (e.g., "aesthetic value" or "goodness"). In the case of identity matching, the relation is one of sameness or similarity, but there are many other types of non-arbitrary relational responding as well. For instance, in non-arbitrary comparative relational responding, the subject must respond by consistently choosing a comparison stimulus either bigger or smaller than the sample.

Arbitrarily Applicable Relational Responding

A range of different species including, for example, humans (e.g., Reese, 1961), monkeys (e.g., Harmon, Strong & Pasnak, 1982), pigeons (e.g., Wright & Delius, 1994), fish (Perkins, 1931) and bees (Giurfa, Zhang, Jenett, Menzel & Srinivasan, 2001) have demonstrated non-arbitrary relational responding. However, relational frame theorists argue that typical members of the human species exposed to contingencies provided by their socioverbal community can learn to show an additional type of relational responding referred to as arbitrarily applicable relational responding (AARR). This is primarily based not on non-arbitrary or formal relations between the stimuli being related but on aspects of the context that specify the relation such that the relational response can be brought to bear on any relata regardless of their non-arbitrary properties (see for example, Stewart & McElwee, 2009).

As an example of AARR, imagine I teach a verbally able child that "Mr. X is taller than Mr. Y and Mr. Y is taller than Mr. Z," and imagine that the child is then able to derive multiple new relations including, for instance, "Mr. X is taller than Mr. Z" and "Mr. Z is shorter than Mr. X." The child is able to do this despite the fact that the latter performances have not been explicitly taught and that the non-arbitrary properties of the stimuli involved do not support these performances (for instance, it is not obvious that the stimulus "Mr. X" is physically taller than the stimulus "Mr. Z"). From an RFT perspective, what is happening is that I am presenting the child with a contextual cue (i.e., "taller") that has been previously established in the child's learning history as controlling a particular pattern of generalized relational responding. When that cue is presented, that response pattern can be brought to bear on any arbitrarily chosen set of stimuli no matter what their non-arbitrary properties or the non-arbitrary relations between them such that all of the stimuli are brought into a coherent set of relations with each other. In this particular case, it is brought to bear on the arbitrary stimuli "Mr. X", "Mr. Y", and "Mr. Z".

Arbitrarily applicable relational responding is an archetypal example of a generalized, purely functionally defined operant (see for example Barnes-Holmes and Barnes-Holmes, 2000). Operant classes are defined functionally, not topographically, meaning that the events of which they are composed (e.g., responses) are classified based on their functional effects as opposed to their topography. However, in practice, members of classes often have topographical features in common and thus for practical purposes are described or defined in topographical terms (e.g., lever pressing typically involves physiologically similar paw movements). Nevertheless, in some cases, including AARR, generalized imitation (Baer, Petersen & Sherman, 1967) and novel responding (Pryor, Haag & O'Reilly, 1969), operant classes have few topographical features in common in terms of either the stimuli or the responses involved and hence rely more obviously on their functional definition. These are described as generalized, overarching or purely functionally defined.

The fact that AARR is a purely functionally defined operant has important implications for how it is established. In training this type of operant, what is critical is the provision of multiple exemplars of exposure to differential consequences for responding in accordance with the pattern of functional relations at issue. Ideally, topography should vary to as great an extent as possible across these training exemplars in order to "wash out" irrelevant elements and facilitate abstraction of the key features of the operant involved. In the case of AARR, these features include responding to a group of stimuli in accordance with a particular pattern under the control of one or more arbitrary contextual cues. In order to train an organism to abstract these features, one must expose the organism to exemplars that allow discrimination between the cue and irrelevant elements including the topography of the related objects.

An example of this training is seen in the way in which very young children are taught symmetrical relations between words and objects, which is arguably the first and most fundamental example of generalized contextually controlled relational responding to be learned. In this case, caregivers will often utter the name of an object in the presence of an infant and then reinforce any orienting response that occurs toward the particular object ("hear name A → look at object B"). They will also often present an object to the infant and then model and reinforce an appropriate naming response ("see object B → hear and say name A"). In this way, they explicitly train the child in both directions in the symmetrical pattern. This informal training consistently occurs in the presence of particular contextual cues such as the phrases "What's that?" and the juxtaposition of words and objects. In addition, it occurs with a wide variety of stimulus objects. Eventually after a sufficient number of exemplars, the generalized response pattern of object-word symmetry is abstracted away from the topography of particular objects and brought under the control of the contextual cues so that derived symmetry (i.e., being able to derive the untaught response when trained in only one direction) with any new word-object pair becomes possible. For instance, imagine a child with this history hears "What's that? That's a [novel name]" (object A → name B) in the presence of a novel object. Contextual cues such as those suggested will now be discriminative for symmetrical responding between the name and the object so that, without any additional training, the child is likely to point to the novel object when asked "Where's the [novel name]?" (name B → object A) and to answer "[novel name]" when presented with the novel object and asked "What's that?" (object B → name A).

Thus, symmetrical responding between words and objects is seen as a generalized or overarching response class generated by a history of reinforcement across multiple exemplars, and once this class has been acquired, the child can derive an untaught symmetrical relation from a trained relation no matter what the physical features of the word-object pair involved. As suggested, this is likely the first example of generalized, contextually controlled relational responding to be established and it is a fundamentally important one since it underlies linguistic reference. The word-object relation thus established can also be seen as a preliminary version of an arbitrarily applicable relation of sameness or coordination, since the child has been taught to treat a word and its referent as functionally similar in particular contexts. Furthermore, RFT suggests that on the basis of continuing exposure to socioverbal interactions, the child will learn to respond in accordance with relations of sameness that involve more than two stimuli as well as relations of non-sameness such as comparison, distinction and opposition.

Learning to respond in accordance with patterns of contextually controlled sameness involving more than two stimuli is, from an RFT point of view, what allows the phenomenon of stimulus equivalence. Sidman suggested that the latter involved three properties: reflexivity, symmetry and transitivity. The first of these, reflexivity, is another name for identity matching, or non-arbitrary sameness responding, which is a simpler type of relational responding than AARR in accordance with sameness and thus highly probable to be present given the latter. The other two properties, symmetry and transitivity, both fall out of the overarching pattern of AARR sameness responding. We have just described the kind of training that enables generalized word-object symmetry and suggest that the transitivity of AARR in accordance with sameness is similarly learned through multiple exemplar training. For example, the child may learn transitive responding in the presence of cues such as "goes with" based on multiple exemplars in which if X (e.g., spoken word "dog") "goes with" Y (e.g., picture of dog) and Y "goes with" Z (e.g., written word "dog") then X (spoken word) "goes with" Z (picture) and vice versa. In addition, the matching-to-sample format, which is used to probe for stimulus equivalence, likely becomes a contextual cue for AARR in accordance with sameness early on as well since this format is one which is employed in many early educational contexts. In a typical exercise using this format, children are required to look at a word or picture and then to choose an appropriate corresponding word or picture from an array, perhaps accompanied by a prompt to choose the one that "matches" or "goes with" the first one. Thus, many humans have a history that has established AARR in accordance with sameness in the context of very early language training and in addition has established the MTS format as a particularly potent cue for this pattern of responding. From an RFT point of view, this can explain the phenomenon of stimulus equivalence itself, as well as the link between stimulus equivalence and language.

Arbitrarily applicable relational responding in accordance with sameness or coordination is the first pattern of AARR to emerge, and there is empirical research with infants charting the emergence of this pattern as well as demonstrating how it can be explicitly trained (Lipkens, Hayes & Hayes, 1993; Luciano, Gomez-Becerra, & Rodriguez-Valverde, 2007). Many other patterns of AARR are learned also. As an example, consider how AARR in accordance with comparison might be acquired. In this case, the child likely first learns to choose the physically larger of two objects in the presence of auditory stimuli such as "bigger" and to choose the physically smaller of two objects in the presence of stimuli such as "smaller". Then through exposure to multiple exemplars of this type of pattern in the presence of these contextual cues, the relational response becomes abstracted such that it can be applied in conditions in which there is no obvious formal relation; for example, after being told that "Mr. A is bigger than Mr. B," a child will be able to derive that "Mr. B is smaller than Mr. A."

Patterns of arbitrarily applicable relational responding are referred to as relational frames. RFT suggests the existence of a variety of these frames including sameness (or coordination, e.g., "A = B"), comparison (e.g., "C is bigger than D"), opposition ("black is the opposite of white"), distinction ("this is not the same as that"), hierarchy ("a whale is a type of mammal"), analogy ("A is to B as C is to D"), deixis ("I am here and you are there"), and temporality ("spring comes before summer") amongst others. Furthermore, RFT researchers have provided an increasing quantity of empirical evidence both that people respond in accordance with a variety of these frames and that they can be trained up when weak or absent (e.g., Barnes-Holmes, Barnes-Holmes, Smeets, Strand & Friman, 2004; Barnes-Holmes, Barnes-Holmes & Smeets, 2004; Berens & Hayes, 2007; Carpentier, Smeets & Barnes-Holmes, 2003; McHugh,

Barnes-Holmes, Barnes-Holmes & Stewart, 2006; Roche & Barnes, 1997; Rosales, Rehfeldt & Lovett, 2011; Steele & Hayes, 1991; see also Dymond, May, Munnelly & Hoon, 2010; Rehfeldt & Barnes-Holmes, 2009).

Properties of Relational Framing

Despite the variety of relational frames, they all share three core properties which are mutual entailment, combinatorial entailment and transformation of function. *Mutual entailment* describes that feature of relational framing whereby a unidirectional relation from stimulus A to stimulus B in a certain context entails a second unidirectional relation from B to A. For instance, imagine a child is shown two previously unseen identically sized foreign coins and told that coin A is worth more than coin B. If she subsequently derives that coin B is less than coin A, then this is mutual entailment. Note that symmetry, which we have discussed as the first type of generalized contextually controlled relational responding to be learned, and which Sidman described as one of the three properties of stimulus equivalence, is a sub-type of mutual entailment in which the entailed relation is the same as the specified relation. In the context of equivalence, for example, if "A goes with B" then "B goes with A."

Combinatorial entailment is the phenomenon whereby two stimulus relations can be combined to allow the derivation of a third relation. For example, imagine that I show the same child as in the example above three identically sized foreign coins and tell her that coin A is worth more than coin B and coin B is worth more than coin C. If she subsequently derives that A is worth more than C and C is worth less than A, then this is combinatorial entailment. Transitivity, another of the properties of equivalence, is a sub-type of combinatorial entailment in which the entailed relation is the same as the specified relation. In the context of equivalence, for example, if "A goes with B" and "B goes with C" then "A goes with C" and "C goes with A."

Transformation of functions is extremely important in terms of the psychological relevance of relational framing since, from an RFT perspective, it is the process by which language can influence our behavior (e.g., Dymond & Rehfeldt, 2000). If two arbitrary stimuli, A and B, participate in a relation and stimulus A has a psychological function, then under certain conditions the stimulus functions of B may be transformed in accordance with that relation. For instance, imagine that the child in the above exercises and I are in a foreign country in which coins that the child has never seen before can be used to buy things (in other words, they now have an appetitive function), and imagine that I show her a previously unseen coin A and ask if she wants this coin. She is likely to reply that she does. Imagine that I then show her a second novel coin, B. If I then tell her that A is worth less than B and then ask her which she wants, she will likely choose B, whereas if I tell her that A is worth more than B then she will likely choose A. In this way, relational framing with respect to stimuli in our environment can transform the psychological functions of those stimuli. Transfer of functions, which was discussed in the context of equivalence, is a subtype of transformation of functions in which the psychological function that appears in the related stimulus is the same as the function inhering in the original stimulus.

The phenomenon of transformation of functions through arbitrarily applicable relations has by now been demonstrated in well over a hundred RFT studies with a variety of different relations (e.g., coordination, distinction, opposition, comparison, analogy, temporality, perspective) and functions (e.g., elicited, conditioned elicited, discriminative, reinforcing, ordering, self-discriminative, avoidance) (see Dymond et al., 2010 for an overview). From an RFT perspective, this phenomenon allows a technical understanding of the influence of language over our behavior and the generativity and flexibility that characterize this influence. It allows insight into rule following, for example, one of the distinctly human phenomena that had posed problems for the Skinnerian conception of language.

RFT provides an analysis of rule-governed behavior in terms of the relational frames involved and the cues that occasion the derivation of those relations (referred to as Crel cues), and also in terms of the psychological functions transformed through those relations and the cues that occasion those transformations of function (referred to as Cfunc cues). Consider the following rule, which might be explicitly provided by the experimenter in a study on rule-governed behavior: "If you press fast when the blue light is on and press slow when the red light is on, then you will earn the maximum number of points, which are exchangeable for money afterwards." From the RFT perspective, an analysis of this rule points to the following relational frames: coordination between words (e.g., blue light) and actual objects or events; before-after relations specified in terms that indicate a temporal antecedent (e.g., "When the blue light is on"); perspective (I-you) relations; and if-then relations that specify a contingency. As regards transformation of functions, the words "press fast" and "press slow" are Cfunc stimuli that alter the behavioral functions of the button in an operant laboratory such that the listener is more likely to press it at a particular rate in particular contexts specified in the rule (e.g., "When the blue light is on" and "When the red light is on" respectively). RFT suggests that someone provided with a rule such as this determines whether or not the rule is being followed by the extent to which the rule coordinates with actual behavior. More technically, for the rule follower, the coordination between the relational network constituted by the rule and the relations sustained among the objects or events specified by the rule acts as an ongoing source of behavioral regulation. In other words, if the listener discriminates that events in the non-arbitrary environment specified by the rule are indeed in the relations specified by the rule, then the rule is being followed. In the above case, for example, if the listener perceives that he is pressing fast when the blue light is on, then he is following the rule correctly.

Thus, RFT provides for the theoretical analysis of rule-governed behavior and how it might affect human behavior. In the case just given, the transformation of the functions of aspects of the operant set-up through relational frames as just described might allow immediate adjustment to experimental contingencies. Furthermore, relational frame researchers have already begun to empirically model simple examples of rule-governed behavior in the laboratory by establishing contextual cues for sameness, distinction and temporal (before-after) relations so as to allow transformation of the ordering functions of arbitrary stimuli and, more specifically, influence over patterns of sequential responding to those stimuli (e.g., O'Hora, Barnes-Holmes, Roche, & Smeets, 2004). Work such as this on rules and other examples of complex behavior is allowing behavior analytic insight into the potentially profound effects of human verbal behavior and the processes through which it can produce such effects (see, for example, Torneke, Luciano & Valdivia Salas, 2008).

Relational Framing and Language

Relational Frame Theory thus constitutes a modern behavior analytic approach to human language. From an RFT perspective, relational framing is the process that underlies language (see e.g., Hayes et al., 2001). In contrast to the Skinnerian approach, the RFT account is a truly functional analytic one, since the behavior of interest is explained in terms of the history of a single organism. The history required involves being trained to show patterns of relational responding based on aspects of the context that specify the relation such that the relational response can be brought to bear on any relata regardless of physical properties, referred to as relational framing. Participants in operant experiments cannot be said to be showing verbal responding unless they demonstrate patterns of behavior consistent with such a history. Individuals with a sufficiently advanced repertoire of this form of behavior, on the other hand, can both speak with meaning, thus producing sequences of stimuli that may come to influence other verbal individuals, and listen with understanding with respect to the stimuli produced by the latter.

Relational framing is potentially extremely generative. Any stimulus can be related to any other stimulus in accordance with any relational frame, including those listed above as well as others. Of course, since natural language has its greatest use in the description and analysis of the formal features of our environment, the fact that relational framing is arbitrarily applicable does not mean that it is arbitrarily applied. Nevertheless, the empirically established and analyzable process of relational framing can explain the generativity and flexibility that language facilitates, from trivial everyday examples (e.g., naming a pet dog) to the extremes of artistic creativity (e.g., writing a best-selling novel) or scientific innovation (e.g., developing a comprehensive empirically based theory of the origins of the universe).

From this perspective, any objects or events that are relationally framed become verbal for us—they become part of the world as known through relational frames. As we frame objects, events and people through our interactions with the socioverbal community, we elaborate our network of related stimuli, and through transformation of functions, the world increasingly takes on new verbally derived functions. The expansion of this network starts when we first learn to frame words and objects as the same and probably continues throughout most of our lives.

The well-documented "language explosion" between the ages of 2 and 3 is an obvious and salient example of the elaboration of the relational network. This typically occurs around the time that children have likely acquired the ability to frame in accordance with a few simple relations, allowing them to derive multiple novel relations amongst an expanding set of named objects and events. As children grow into adulthood, continued verbal interactions produce an increasingly complex and multi-relational network involving vast numbers of different objects and events and the relations between them. For human beings, everything we encounter and think about, including ourselves, our thoughts and emotions, our prospects, other people, and our environment, becomes part of this elaborate verbal relational network. Thus, for human beings the world is verbal, and we can never completely escape language except under very unusual circumstances.

The fact that events and objects in our world take on verbal functions through relational framing has both positive and negative ramifications for us (see Torneke, 2010). On the one hand, the human ability to verbally analyze and transform the functions of our environment can allow more effective behavior, ultimately including science and technology (Hayes, Gifford, Townsend & Barnes-Holmes, 2001). However, the very same capacity can sometimes cut individuals off from important experiences and produce maladaptive responding (see, e.g., Hayes, Strosahl & Wilson, 2011). In any event, with regard to both negative and positive aspects, the emergence of a scientifically adequate theory that allows us to analyze the basic processes at work represents an important advance in both basic and applied domains.

As we have seen, RFT can readily explain both of the empirical phenomena that posed problems for the Skinnerian account. In addition, it was suggested that a key sign of a scientifically adequate account of an important psychological phenomenon is that the account could result in a progressive program of research into that phenomenon, giving new insight into psychologically important processes and raising new research questions as it does so. Skinner's account has arguably failed in this key respect since more than half a century after the publication of verbal behavior there is little or no empirical work investigating key phenomena of human language or cognition that is primarily based on the Skinnerian approach (Dymond, & Alonso-Álvarez, 2010).

In contrast, RFT has given rise to a vibrant program of research that has seen advances on a number of fronts, particularly over the last decade (see Dymond, et al., 2010). The research that has been conducted has involved the investigation of aspects of the RFT account itself as well as the use of this conception of language to explore important linguistic phenomena. We have already considered rule-governed behavior. Other areas of exploration include generative verbal behavior in developmentally delayed children (e.g., Moran, Stewart, McElwee & Ming, 2010; Murphy & Barnes-Holmes 2009; Heagle & Rehfeldt, 2006), analogical reasoning in typically developing children and adults (Carpentier, Smeets & Barnes-Holmes, 2003; Stewart, Barnes-Holmes & Roche, 2004; Stewart, Barnes-Holmes & Weil, 2009), development of novel methodologies for examining derived relations (e.g., Dymond & Whelan, 2010; O'Hora, et al., 2004; Smyth, Barnes-Holmes & Forsyth, 2006; Stewart, Barnes-Holmes & Roche, 2004), assessment and training of intelligent performance (e.g., O'Toole & Barnes-Holmes, 2009; Cassidy, Roche & Hayes, 2011), implicit cognition and measurement of socially sensitive attitudes (Barnes-Holmes, Barnes-Holmes, Stewart & Boles, 2010), the assessment of perspective taking in applied developmental and clinical arenas (McHugh, Barnes-Holmes & Barnes-Holmes, 2004; Rehfeldt, Dillen, Ziomek & Kowalchuk, 2007; Villatte, Monestes, McHugh, Freixa i Baque & Loas, 2010a & b; Weil, Hayes, & Capurro, 2011) and verbal processes underlying psychopathology (e.g., Dymond & Roche, 2009; Gannon, Roche, Kanter, Forsyth & Linehan, 2011; Roche, Kanter, Browne, Dymond & Fogarty, 2008).

One of the primary roles of this volume is to provide an update on the Relational Frame Theory based research work that has been occurring over the past decade since the publication of the first book on RFT in 2001. The research areas that have just been listed are all areas that researchers have explored over the last decade and that are discussed in detail in the present volume. In this work the promise of RFT as a scientifically adequate account of language is beginning to bear fruit.

References

Baer, D. M., Petersen, R. F., & Sherman, J. A. (1967). The development of imitation by reinforcing behavioral similarity to a model. *Journal of the Experimental Analysis of Behavior, 10*, 405-416.

Barnes, D., McCullagh, P., & Keenan, M. (1990). Equivalence class formation in non-hearing impaired children and hearing impaired children. *Analysis of Verbal Behavior, 8*, 1-11.

Barnes-Holmes, D., & Barnes-Holmes, Y. (2000). Explaining complex behavior: Two perspectives on the concept of generalized operant classes? *The Psychological Record, 50*, 251-265.

Barnes-Holmes, Y., Barnes-Holmes, D., Smeets, P. M. (2004). Establishing relational responding in accordance with opposite as generalized operant behavior in young children. *International Journal of Psychology and Psychological Therapy, 4*, 559-586.

Barnes-Holmes, Y., Barnes-Holmes, D., Smeets, P. M., Strand, P., & Friman, P. (2004). Establishing relational responding in accordance with more-than and less-than as generalized operant behavior in young children. *International Journal of Psychology and Psychological Therapy, 4*, 531-558.

Barnes-Holmes, D., Barnes-Holmes, Y., Stewart, I., & Boles, S. (2010). A sketch of the Implicit Relational Assessment Procedure (IRAP) and the Relational Elaboration and Coherence (REC) model. *The Psychological Record, 60*, 527-542.

Berens, N. M., & Hayes, S. C. (2007). Arbitrarily applicable comparative relations: Experimental evidence for relational operants. *Journal of Applied Behavior Analysis, 40*, 45-71.

Carpentier, F., Smeets, P. M., & Barnes-Holmes, D. (2003). Equivalence-equivalence as a model of analogy: Further analyses. *The Psychological Record, 53*, 349-372.

Cassidy, S., Roche, B., & Hayes, S. C. (2011). A relational frame training intervention to raise Intelligence Quotients: A pilot study. *The Psychological Record, 61*, 173-198.

Dougher, M. J., Augustson, E., Markham, M. R., Greenway, D. E., & Wulfert, E. (1994). The transfer of respondent eliciting and extinction functions through stimulus equivalence classes. *Journal of the Experimental Analysis of Behavior, 62*, 331-351.

Dymond, S., & Alonso-Álvarez, B. (2010). The selective impact of Skinner's *Verbal Behavior* on empirical research: A reply to Schlinger (2008). *The Psychological Record, 60*, 355-360.

Dymond, S., May, R. J., Munnelly, A., & Hoon, A. E. (2010). Evaluating the evidence base for relational frame theory: A citation analysis. *The Behavior Analyst, 33*, 97-117.

Dymond, S., & Rehfeldt, R. A. (2000). Understanding complex behavior: The transformation of stimulus functions. *The Behavior Analyst, 23*, 239-254.

Dymond, S., & Roche, B. (2009). A contemporary behavior analysis of anxiety and avoidance. *The Behavior Analyst, 32,* 7-28.

Dymond, S., Roche, B., & Barnes-Holmes, D. (2003). The continuity strategy, human behavior, and behavior analysis. *The Psychological Record, 53,* 333-347.

Dymond, S., & Whelan, R. (2010). Derived relational responding: A comparison of matching to sample and the relational completion procedure. *Journal of the Experimental Analysis of Behavior, 94,* 37-55.

Galizio, M. (1979). Contingency-shaped and rule-governed behavior: Instructional control of human loss avoidance. *Journal of the Experimental Analysis of Behavior, 31,* 53-70.

Gannon, S., Roche, B., Kanter, J., Forsyth, J. P., & Linehan, C. (2011). A derived relations analysis of approach-avoidance conflict: Implications for the behavioral analysis of human anxiety. *The Psychological Record, 61(2),* 227-252.

Giurfa, M., Zhang, S., Jenett, A., Menzel, R., & Srinivasan, M. V. (2001). The concepts of sameness and difference in an insect. *Nature, 410,* 930-932.

Gross, A. C., & Fox, E. J. (2009). Relational frame theory: An overview of the controversy. *The Analysis of Verbal Behavior, 25,* 87-98.

Harmon, K., Strong, R., & Pasnak, R. (1982). Relational responses in tests of transposition with rhesus monkeys. *Learning and Motivation, 13 (4),* 495-504.

Hayes, S. C. (Ed.). (1989). *Rule-governed behavior: Cognition, contingencies, and instructional control.* New York: Plenum.

Hayes, S. C., Fox, E., Gifford, E. V., Wilson, K. G., Barnes-Holmes, D. & Healy, O. (2001). Derived relational responding as learned behavior. In S. C. Hayes, D. Barnes-Holmes & B. Roche (Eds.) *Relational frame theory: A post-Skinnerian account of human language and cognition.* New York: Plenum Press.

Hayes, S. C., Gifford, E. V., Townsend, R. C., & Barnes-Holmes, D. (2001). Thinking, problem-solving and pragmatic verbal analysis. In S. C. Hayes, D. Barnes-Holmes & B. Roche (Eds.) *Relational frame theory: A post-Skinnerian account of human language and cognition.* New York: Plenum Press.

Hayes, S. C., Strosahl, K., & Wilson, K. G. (2011). *Acceptance and Commitment Therapy: The process and practice of mindful change (2nd ed.).* New York: Guilford Press.

Heagle, A. I., & Rehfeldt, R. A. (2006). Teaching perspective-taking skills to typically developing children through derived relational responding. *Journal of Early and Intensive Behavioral Intervention, 3,* 1-34.

Lipkens, R., Hayes, S. C., & Hayes, L. J. (1993). Longitudinal study of the development of derived relations in an infant. *Journal of Experimental Child Psychology, 56,* 201-239.

Luciano, C., Gomez Becerra, I., & Rodriguez Valverde, M. (2007). The role of multiple-exemplar training and naming in establishing derived equivalence in an infant. *Journal of the Experimental Analysis of Behavior, 87,* 349-365.

McHugh, L., Barnes-Holmes, Y., Barnes-Holmes, D. (2004). Perspective-taking as relational responding: A developmental profile. *The Psychological Record, 54,* 115-144.

McHugh, L., Barnes-Holmes, Y., Barnes-Holmes, D., & Stewart, I. (2006). Understanding false belief as generalized operant behavior. *The Psychological Record, 56*, 341-364.

Moran, L., Stewart, I., McElwee, J., & Ming, S. (2010). The Training and Assessment of Relational Precursors and Abilities (TARPA): A preliminary analysis. *Journal of Autism and Developmental Disorders, 40(9)*, 1149-1153.

Murphy, C., & Barnes-Holmes, D. (2009). Establishing derived manding for specific amounts with three children: An attempt at synthesizing Skinner's *Verbal Behavior* with Relational Frame Theory. *The Psychological Record, 59*, 75-92.

Ogawa, A., Yamazaki, Y., Ueno, K., Cheng, K. & Iriki, A. (2009). Neural correlates of species-typical illogical cognitive bias in human inference. *Journal of Cognitive Neuroscience, 22(9)*, 2120-2130.

O'Hora, D., & Barnes-Holmes, D. (2001). Stepping up to the challenge of complex human behavior: A response to Ribes-Inesta's response. *Behavior and Philosophy, 29*, 59-60.

O'Hora, D., Barnes-Holmes, D., Roche, B., & Smeets, P. M. (2004). Derived relational networks and control by novel instructions: A possible model of generative verbal responding. *The Psychological Record, 54*, 437-460.

O'Toole, C., & Barnes-Holmes, D. (2009). Three chronometric indices of relational responding as predictors of performance on a brief intelligence test: The importance of relational flexibility. *The Psychological Record, 59*, 119-132.

Perkins, F. T. (1931). A further study of configurational learning in the goldfish. *Journal of Experimental Psychology, 14*, 508-538.

Pryor, K. W., Haag, R., & O'Reilly, J. (1969). The creative porpoise: Training for novel behavior. *Journal of the Experimental Analysis of Behavior, 12*, 653-661.

Reese, H. W. (1961). Transposition in the intermediate size problem by pre-school children. *Child Development, 32*, 311-314.

Reese, H. W. (1968). *The perception of stimulus relations: Discrimination learning and transposition*. Academic Press: New York.

Rehfeldt, R. A., & Barnes-Holmes, Y. (Eds.) (2009). *Derived relational responding: Applications for learners with autism and other developmental disabilities*. Oakland, CA: Context Press/New Harbinger.

Rehfeldt, R. A., Dillen, J. E., Ziomek, M. M., & Kowalchuk, R. K. (2007). Assessing relational learning deficits in perspective-taking in children with high functioning autism spectrum disorder. *The Psychological Record, 57*, 23-47.

Roche, B., & Barnes, D. (1997). A transformation of respondently conditioned stimulus function in accordance with arbitrarily applicable relations. *Journal of the Experimental Analysis of Behavior, 67*, 275-300.

Roche, B., Kanter, J. W., Brown, K., Dymond, S., & Fogarty, C. (2008). A comparison of "direct" versus "derived" extinction of avoidance. *The Psychological Record, 58*, 443-464.

Rosales, R., Rehfeldt, R. A., & Lovett, S. (2011). Effects of multiple exemplar training on the emergence of derived relations in preschool children learning a second language. *The Analysis of Verbal Behavior, 27*, 61-74.

Saunders, K., & Williams, D. (1998). Stimulus control procedures. In K. A. Lattal & M. Perone (Eds.) *The handbook of research methods in human operant behavior* (pp. 193-228). New York: Plenum Press.

Sidman, M. (1971). Reading and auditory-visual equivalences. *Journal of Speech and Hearing Research, 14,* 5-13.

Skinner B. F. (1957). *Verbal behavior.* New York: Appleton-Century-Crofts.

Skinner, B. F. (1969). *Contingencies of reinforcement: A theoretical analysis.* Appleton-Century-Crofts: New York.

Smyth, S., Barnes-Holmes, D., & Forsyth, J. P. (2006). A derived transfer of simple discrimination and self-reported arousal functions in spider-fearful and non-spider-fearful participants. *Journal of the Experimental Analysis of Behavior, 85,* 223-246.

Steele, D. L., & Hayes, S. C. (1991). Stimulus equivalence and arbitrarily applicable relational responding. *Journal of the Experimental Analysis of Behavior, 56,* 519-555.

Stewart, I., Barnes-Holmes, D., & Roche, B. (2004). A functional-analytic model of analogy using the relational evaluation procedure. *The Psychological Record, 54,* 531-552.

Stewart, I., Barnes-Holmes, D., & Weil, T. (2009). Training analogical reasoning as relational responding. In R. A. Rehfeldt & Y. Barnes-Holmes (Eds.), *Derived relational responding: Applications for learners with autism and other developmental disabilities* (pp. 257-279). Oakland, CA: Context Press/New Harbinger.

Stewart, I., & McElwee, J. (2009). Relational responding and conditional discrimination procedures: An apparent inconsistency and clarification. *The Behavior Analyst, 32(2).* 309-317.

Torneke, N. (2010). *Learning RFT.* Oakland, CA: New Harbinger.

Torneke, N., Luciano, C., & Valdivia Salas, S. (2008). Rule-governed behavior and psychological problems. *International Journal of Psychology and Psychological Therapy, 8(2),* 141-156.

Villatte, M., Monestes, J. L., McHugh, L., Freixa i Baque, E., & Loas, G. (2010a). Adopting the perspective of another in belief attribution: Contribution of Relational Frame Theory to the understanding of impairments in schizophrenia. *Journal of Behavior Therapy and Experimental Psychiatry, 41,* 125-134.

Villatte, M., Monestes, J. L., McHugh, L., Freixa i Baque, E., & Loas, G. (2010b). Assessing perspective taking in schizophrenia using Relational Frame Theory. *The Psychological Record, 60,* 413-424.

Weil, T. M., Hayes, S. C., & Capurro, P. (2011). Establishing a deictic relational repertoire in young children. *The Psychological Record, 61,* 371-390.

Wright, A. A., & Delius, J. D. (1994). Scratch and match: Pigeons learn matching and oddity with gravel stimuli. *Journal of Experimental Psychology: Animal Behavior Processes, 20,* 108-112.

CHAPTER FOUR

Reframing Relational Frame Theory Research: Gaining a New Perspective through the Application of Novel Behavioral and Neurophysiological Methods

Robert Whelan

University of Vermont

Michael W. Schlund

*Kennedy Krieger Institute,
Johns Hopkins University School of Medicine,
& University of North Texas*

As the contents of the present volume attest, Relational Frame Theory (RFT) is an ambitious attempt to explain many aspects of complex human behavior, such as language, deductive reasoning, and perspective taking. Adequately characterizing complex behavior with precision, scope and depth will require the use of innovative procedures to train and test for derived relations—and increasingly sensitive methods to measure such behavior. Derived relational responding (DRR) has typically been studied using a matching-to-sample (MTS) preparation, with accuracy as the dependent variable. However, MTS, while suitable for behavior-analytic approaches that only consider the relation of equivalence (e.g., Horne & Lowe, 1996; Sidman, 1994) does not easily lend itself to the study of more complex forms of derived relations. Indeed, a reliance on MTS may itself narrow the focus of the research agenda, by promoting class-based analyses (Hayes & Barnes, 1997).

A key feature of new protocols developed as alternatives to MTS is that a potentially infinite number of relational responses can be trained and tested. In addition, the use of accuracy as a dependent measure is best suited to classifying responses into two categories—correct or incorrect. Complex human behavior, such as language, needs measures that can capture subtleties in responding (e.g., response accuracy cannot tell us how robins, penguins and ostriches differ along a continuum of "bird-like" qualities). Alternatives to accuracy as a dependent variable include reaction time (RT) and measures of brain activity such as electroencephalography (EEG), event-related potentials (ERPs) and blood-oxygen-level-dependent (BOLD) functional magnetic resonance imaging (fMRI). This chapter argues that these measures have the ability to open up new, fruitful avenues of RFT research, yielding novel results that cannot be found using traditional methods. First, we begin with a review of novel procedures in DRR research, followed by an overview of novel applications of behavioral measures alone and with neurophysiological measures.

Overview of Procedures to Study Derived Relational Responding

Lipkens and Hayes (2009) commented that excessive use of MTS procedures in RFT research may constrain the development of the experimental analysis of DRR. Consequently, the explanatory power of RFT accounts of complex human behaviors would be severely limited because of concerns largely based on the face and ecological validity of the methods. For example, the MTS format is rarely found in real world learning and therefore research based around MTS procedures may be difficult for some to extrapolate from the laboratory to everyday events. In addition, MTS protocols may be best suited for studies of equivalence relations, thereby limiting investigation of multiple stimulus relations such as opposition or comparative relations (see Hayes & Barnes, 1997 for a discussion). MTS protocols often involve extensive training and testing, and this both is inefficient and potentially introduces confounding variables (Horne & Lowe, 1997). Given these concerns and limitations, researchers and clinicians alike have begun to explore other more flexible and efficient protocols and measures, and in doing so they have gained new perspectives on RFT and DRR.

Variants of MTS

The Relational Evaluation Procedure (REP), and its precursor, the pREP, has been employed to study Same and Different (Stewart, Barnes-Holmes & Roche, 2004), Before and After (O'Hora, Barnes-Holmes, Roche, & Smeets, 2004; see also Cullinan, Barnes-Holmes, & Smeets, 2001; Lipkens & Hayes, 2009), and analogical relations (Stewart et al., 2002; Stewart et al., 2004). The key feature of the REP is that it allows participants to evaluate, or report on, different stimulus relations by confirming or denying the applicability of particular stimulus relations to other sets of stimulus relations. For example, a participant might be presented with a contextual cue for SAME and two or more arbitrary stimuli that are specified within that trial as participating in a Same relation. The participant is then required to select one of two arbitrary shapes that function as True or False. For instance, selecting True in the presence of the SAME contextual cue, and selecting False in the presence of an OPPOSITE contextual cue would be considered correct responses. In contrast to MTS protocols, once appropriate contextual control is acquired in the REP, the types of relations that can be generated are not constrained by the prior training and testing of a specific set of derived relations.

The Relational Completion Procedure (RCP; Dymond, Roche, Forsyth, Whelan, & Rhoden, 2007, 2008; Dymond & Whelan, 2010) is a procedure based on the REP that has been used to train and test Same and Opposite relations. The RCP was originally developed to provide a functional analytic model of reading and responding to sentence-completion tasks. The RCP, therefore, involves presenting stimuli in sequence from left to right (as English is read left-to-right), starting with the sample and followed by a contextual cue and a blank space in the upper portion of the screen. The comparison stimuli then appear in the lower portion of the screen. The participants' task is to "complete the sentence" by dragging and dropping one of several comparisons into the blank space and confirming each selection before proceeding to the next trial (the trial can be re-started by the participant if required). For example, if given a large square as sample, a contextual cue for OPPOSITE, then dragging the smallest square among the five squares of different sizes and emitting the confirmatory response would be reinforced. Selection-based feedback is also presented. That is, the sample, contextual cue, and selected comparison are presented along with the feedback ("correct" or "wrong"). During arbitrary relational training and testing, the response format is identical except for the use of arbitrary stimuli, such as nonsense syllables.

Across Dymond et al. (2007, 2008), 22 out of 25 participants passed the arbitrary relational test, considerably better than when MTS-based procedures have been used to study Same and Opposite relations (e.g., Whelan & Barnes-Holmes, 2004). Dymond and Whelan (2010) systematically varied some parameters of the RCP and MTS protocols, including the number of comparisons (three versus five) and the presence or absence of the confirmatory response. The RCP was generally slightly better than the MTS in terms of yield and cycles to criterion. However, the findings clearly identified the confirmatory response requirement as perhaps the critical component of both tasks, possibly because the relation on the screen is likely to be evaluated in a similar manner to the "True" and "False" functions of the REP. That is, in the RCP, participants first complete the relation and then evaluate it, while in the MTS, participants select the relation and then evaluate it. Importantly, the REP and RCP procedure can accommodate all combinations of multiple stimulus relations, which make them flexible alternatives to MTS.

Gaining Insights from Novel Behavioral Measures of Derived Relational Responding

Response accuracy is the most common measure of derived relations. Thus, *proportion of test trials correct* is the measure by which the emergence of derived relations is typically inferred (Dymond & Rehfeldt, 2001). The criterion for accurate responding for normally developing human subjects is typically above 90% correct responding on the critical test trials (e.g., Wang, Dack, McHugh, & Whelan, 2011) or higher in specific populations (e.g., 100% with children with autism; Murphy, Barnes-Holmes, & Barnes-Holmes, 2005). It is important to remember that, from a behavior-analytic perspective, the emergence of derived relations and a test score of 90% accuracy or higher are not one and the same; the latter is merely the most frequently adopted measure for studying the former (Dymond & Rehfeldt, 2001). In addition, accuracy may not be a particularly sensitive measure of DRR (e.g., detecting differences among types of derived relations, such as mutual entailment versus combinatorial entailment).

Reaction Time

Reaction or response time (RT), widely employed as a dependent measure in other areas of psychology (see Posner, 2005, for a review), has provided a novel perspective of DRR that is not possible with accuracy measures provided care is taken when analyzing RT data (Whelan, 2008). Furthermore, employing RT as a dependent measure of DRR provides a means of linking RFT research to a host of issues more commonly addressed in cognitive psychology—thus, it provides a form of common currency. For example, network theories of verbal or semantic meaning (e.g., Collins & Loftus, 1975) appear to overlap conceptually with DRR (Cullinan, Barnes, Hampson, & Lyddy, 1994). Therefore, findings observed using semantic stimuli should also be found when using stimuli from derived relational networks. *Semantic priming* is typically observed during a *lexical decision task* and refers to an improvement in RT or accuracy to a target stimulus when it is preceded by a semantically related prime stimulus (e.g., cat-dog), relative to when it is preceded by a semantically unrelated prime stimulus (e.g., table-dog). Hayes and Bissett (1998) provided a first demonstration of priming with intra-experimentally defined relational networks derived by showing that RTs were faster to pairs of stimuli in a derived relation versus unrelated pairs.

Barnes-Holmes, Staunton et al. (2005) replicated, and then extended, the priming effects of Hayes and Bissett (1998). Barnes-Holmes, Staunton et al. (2005) employed the more common single-word priming task rather than the two-word task adopted by Hayes and Bissett, who did not present corrective feedback during the lexical decision task. Importantly, in Barnes-Holmes, Staunton et al.'s Experiment 2 the lexical decision task was undertaken *before* participants were exposed to the equivalence test, whereas in Hayes and Bissett all participants were exposed to the equivalence test before exposure to the lexical decision task. Thus, in Hayes and Bissett the four stimuli contained within each of the two equivalence relations had been matched repeatedly (albeit in the absence of reinforcement). Notably, in Barnes-Holmes, Staunton et al.'s Experiment 2, priming was not observed for those participants who subsequently failed the equivalence test, thus excluding the possibility that simple exposure to a set of interrelated conditional discriminations would be sufficient to produce priming. Also, *mediated priming* (priming that involves an intervening stimulus) was not observed in Experiment 1, but was observed in Experiment 2. Barnes-Holmes, Staunton et al. speculated that mediated priming effects were eliminated due to the pairing of the related stimuli in the test phase. Whelan, Cullinan, O'Donovan, and Rodríguez-Valverde (2005) demonstrated priming across both Same and Opposite relations, with significantly longer RTs to unrelated versus related pairs. Again, participants who did not reach the criterion on the test for derived relations did not show priming effects.

Reaction time measures are also particularly useful when studying *nodal distance*, a node being a stimulus that is linked by training to at least two other stimuli (Fields & Verhave, 1987), and the number of nodes that link any two stimuli in a set of trained conditional relations is described as the *nodal distance* (Sidman, 1994; also known as nodal number). Several studies have reported that RT is a direct function of nodal distance (e.g., Wang, McHugh & Whelan, 2012). Importantly, RT appears to be sensitive to nodal effects, or the relatedness among stimuli in general (Wang et al., 2012), even when response accuracy remains intact. For example, Wang et al. (2012) found that response accuracy did not decrease as nodal distance increased, whereas response speed (inverted RT) did. However, nodal distance effects were not observed during the lexical decision task of Barnes-Holmes, Staunton et al. (2005), suggesting that type of task, or some other variable, may play a key role in whether or not such nodal effects are observed.

Reaction time measures of derived analogical responding have also been conducted (e.g., Barnes-Holmes, Regan et al. 2005). Barnes-Holmes, Regan et al. employed an MTS task to train participants in two sets of related conditional discriminations. A pair of nonsense syllables served as the sample and two pairs of nonsense syllables as comparisons. Each pair of syllables was either from the same derived relation or from two different derived relations. Given a sample that contained two elements from the same derived relation, the correct comparison contained two elements from another same relation. Also, given a sample that contained two elements from different derived relations the correct comparison contained two elements from another two different derived relations. Barnes-Holmes, Regan et al. (2005) predicted, and subsequently demonstrated, that RT was significantly faster for similar–similar versus different–different relational responding. Barnes-Holmes, Regan et al. reasoned that this RT difference arose because two separate relational frames (distinction and coordination) were being employed to produce the correct relational response whereas the similar–similar analogical network involves only one type of frame, namely, coordination.

Transitive inference (TI) involves directly training a number of relations (the premises) such as "B is greater than A" and "C is greater than B." Next the derived, non-adjacent, relation is tested (i.e., the inference of the relation): is C greater than A? In the cognitive literature, evidence of this type of comparative relational network is often shown through the *symbolic distance effect* (e.g., Acuna et al., 2002) whereby accuracy increases, and RTs decrease, the further apart stimulus pairs are along the array (cf. nodal distance). The procedures reported in DRR research may provide a means of exploring logical TI abilities in nonhumans without having to provide an extensive language-training history (see Munnelly, Dymond & Hinton, 2010 for a discussion of two types of TI tasks). Briefly, the typical DRR TI task involves a discrimination along a formal dimension (based on training with nonarbitrary stimuli) and subsequent control by the arbitrary contextual cues. In this manner, an n-term series TI task can be trained (e.g., see Whelan, Barnes-Holmes, & Dymond, 2006 for a demonstration of a 7-term linear series).

Reilly, Whelan and Barnes-Holmes (2005) compared RT among pairs of stimuli separated by 1, 2 or 3 non-adjacent pairs and found that the former produced significantly longer latencies than the latter 2 relations, thus demonstrating the symbolic distance effect. Furthermore, the group trained with All-More relations produced significantly faster RTs on mutually entailed trials than a group trained with All-Less relations. Munnelly et al. (2010) partially replicated and extended the findings of Reilly et al. by analyzing both response accuracy and latency to a 5-term TI series in which all of the trained relations were either all More-than relations (the *All-More group*) or all Less-than relations (The *All-Less group*). Most notably, there was a performance advantage on tests that involved the same relation as trained when compared to trials on which the tested relation was different to the trained relation. Reaction times on mutually entailed trials were marginally slower than on directly trained trials, but no significant differences were observed between the All-More and All-Less groups, contrasting with the results of Reilly et al. However, Munnelly et al. noted that participants in Reilly et al. were exposed to the arbitrary relational test up to a maximum of six times, compared to four times in Munnelly et al. It is possible, therefore, that the faster RTs during mutually entailed trials seen by the All-More group relative to the All-Less group in Reilly et al. may have been a function of the repeated training and testing procedures employed (see O'Hora, Roche, Barnes-Holmes, & Smeets, 2002).

Wang et al. (2011) adopted a *transfer-of-function* test as a measure of stimulus relatedness. During a training phase, differential responses were trained, using corrective feedback, to the C and D stimuli in equivalence classes consisting of six members (A-B-C-D-E-F). Next, the A, B, E, and F stimuli were presented in the absence of corrective feedback, and the number of responses to each was recorded. Nodal distance effects were observed, in that the A and B stimuli evoked the response trained to the C stimulus, and the E and F stimuli evoked the response trained to the D stimulus. If the stimuli in the equivalence relations were all equally related then responses to A, B, E, and F should have been distributed equally following the equal reinforcement training. Indeed, the transfer-of-function test was the most sensitive measure of nodal distance effects, response speed (inverted RT) was the next most sensitive measure, and response accuracy was the least sensitive measure. Fields and Watanabe-Rose (2008) speculated that the format of the MTS structure itself occasions responding in accordance with class membership and discrimination between classes. In contrast, the format of the transfer-of-function test is such that responses occur in the presence of members of the same

class and, therefore, occasions responding according to within-class differences, such as nodal distance.

Other Measures

Other measures of DRR have been used in previous research,
although their use appears to have waned in the past decade, despite
their potential utility. Verbal self-reports, usually collected through post-
experimental interviews or questionnaires, have been used as a measure
of equivalence performance (e.g., Arntzen, 2004). However, as Cabello
and O'Hora (2002) note, post-experimental reports assume that verbal
behavior after the experiment accurately reflects the behavior during the
experiment, whereas *protocol analysis*, in which participants concurrently
verbalize their thoughts while they complete an experimental task, may
be more useful. Stimulus sorting tests, in which participants group
stimuli, although widely used in studies on categorization and concept
formation, have infrequently been used in studies on derived stimulus
relations (Smeets, Dymond, & Barnes-Holmes, 2000). Guinther and
Dougher (2010) presented participants with a list of words from within
one of the groups for a free recall test and a recognition test. Participants
were more likely to falsely recall and recognize words that had been
assigned to the same group as the list words during prior training, relative to words not assigned to the same group and relative to words that
co-occurred with list words.

Gaining Insights from Novel Neurophysiological Measures of Derived Relational Responding

There are many tools currently available to the researcher who wishes to
study brain activity: however, EEG and MRI are two of the most accessible tools for studying human brain function *in vivo*. Both EEG and MRI
are noninvasive and can be repeated indefinitely without any known
adverse effects. EEG measures the electrical activity in the brain as
recorded by electrodes on the surface of the scalp which results from the

summed activity of large numbers of pyramidal neurons in the cerebral cortex (i.e., the outer layer of the brain) that have fired simultaneously in the same direction. The recorded scalp potential is both very small (on the order of tens of a millionth of a volt) and spatially distributed (activity at a particular scalp electrode does not necessarily equate to activity in the underlying brain region). Nonetheless, EEG has excellent temporal resolution, on the order of milliseconds.

Functional MRI (fMRI) provides indirect spatiotemporal information about neural activity using blood-oxygenation-level-dependent (BOLD) signal, which is a reflection of changes in the inhomogeneity of the magnetic field resulting from changes in blood oxygenation. Magnet strength, measured in tesla (T), correlates with the ability to detect the BOLD signal, commonly referred to as brain "activation". By contrasting experimental conditions that, at least in principle, differ only in the effect of interest, changes in the magnetic field due to the activity of specific brain regions can be detected. The BOLD signal thus measures brain activity indirectly, and furthermore, the BOLD response is slow (peaking around 4 s), although spatial resolution is on the order of several millimeters. In many respects, EEG and MRI data provide useful and complementary measures of regional responses, but with EEG providing greater temporal resolution and fMRI providing greater spatial resolution.

EEG and ERP Research

Roche, Linehan, Ward, Dymond, and Rehfeldt (2004) exploited the temporal resolution of EEG to demonstrate the operant nature of DRR. The human EEG contains a number of separate "frequency bands", of which *alpha* (8-13 Hz) can be considered as inversely related to the amount of brain resources employed. Roche et al. reported that, in the four subjects for whom adequate data were acquired, alpha band activity reduced as participants' proficiency on Same and Opposite combinatorial relations increased, indicating that fewer brain resources were required to produce the operant response.

Event-Related Potentials (ERPs) are changes in the EEGs that are caused by a particular event (e.g., the presentation of a specific stimulus). At a simplified level, the spontaneous background EEG fluctuations can be considered to occur at random relative to the time point when the stimuli occurred. When EEG is averaged across these time points, the ERPs therefore reflect only that activity which is consistently associated with the stimulus processing. Specific categories of ERP are related to specific kinds of behavior. For example, the N400 ERP (N denoting negative, and 400 the approximate number of ms post-stimulus that the ERP is apparent) is observed when semantically anomalous sentences are presented, such as "They wanted the hotel to look more like a tropical resort. So along the driveway they planted rows of …[palm trees] versus [tulips]." Barnes-Holmes, Staunton et al. (2005), in addition to the RT measures from Experiments 1 and 2 (described above), obtained EEG data in Experiment 3 from 20 participants. The N400 was observed following presentations of unrelated stimulus pairs, and not for related pairs, thus providing a DRR analog of directly related, indirectly related, and unrelated word pairs. The finding that equivalent stimulus pairs produced priming is particularly notable, as the analog effect in the cognitive literature—mediated priming—is typically a weak effect.

In addition to Barnes-Holmes, Staunton et al. (2005), other studies have also examined ERP responses to priming effects with derived relations. Yorio, Tabullo, Wainselboim, Barttfeld, and Segura (2008) recorded EEG data from 10 participants, as these participants completed equivalence tasks. During the test phase, participants were presented with a sample and a single comparison stimulus and were instructed to respond according to whether or not the stimuli were "related" or "unrelated". Rather than the orthodox N400 reported by Barnes-Holmes, Staunton et al., Yorio et al. reported that a positive component from 350 to 600 ms, peaking at 450 ms, was observed over the parietal lobe. This positive component appeared to be a P300 ERP (typically the P300 appears in response to an "oddball"—an infrequent, unexpected stimulus). There are a number of possible reasons that Yorio et al.'s results were different from those of Barnes-Holmes, Staunton et al. (2005). In contrast to the phonologically and orthographically regular (i.e., word-like) stimuli that Barnes-Holmes, Staunton et al. employed, Yorio et al.'s equivalence classes were composed of graphical stimuli. In Yorio et al. the delay between prime and target was 2,500 ms (50 ms in Barnes-Holmes, Staunton et al.), and the target stimulus was terminated only when the participant made a response (a limited hold of 1500 ms was in effect in Barnes-Holmes, Staunton et al.). Haimson, Wilkinson,

Rosenquist, Ouimet, and McIlvane (2009) also employed a lexical decision task. Unfortunately, no statistical testing was employed for the results involving the arbitrary stimuli in equivalence classes and, therefore, it is not possible to say if an N400 ERP was actually produced.

Barnes-Holmes, Regan et al. (2005), in their study of analogy, obtained EEG data in addition to RT data (described above). Significant differences between similar–similar versus different–different responses were found for the left hemisphere scalp regions, beginning approximately 1,000 ms after onset of the comparison stimulus. This difference moved in anterior-posterior direction across the scalp: different–different waveforms became increasingly more negative than similar–similar waveforms. Barnes-Holmes, Regan et al. speculated that this negativity may have been modulated by degree of expectedness of the stimuli. That is, the negative waveforms may have indicated that the analogical network elicited by different–different relational responding was of low probability relative to a similar–similar network. This effect may parallel the relationship between analogies such as "he is to his brother as chalk is to cheese" versus "apple is to orange as dog is to cat."

fMRI Research

The first study to examine the functional neuroanatomy of DRR was conducted by Dickins et al. (2001), who reported data from 12 participants who underwent fMRI at 1.5 T, during MTS tests of directly trained, mutually and combinatorially entailed relations. An identity-matching task served as the control condition. Whole brain analyses of activation during the MTS task showed differences from the verbal fluency task in that no activation of Broca's area (a key region for productive language) was found in any of the MTS tasks and the dorsolateral prefrontal cortex (DLPFC) and posterior parietal cortex activation were more bilateral in nature. Behavioral accuracy for transitive and equivalent relations was significantly correlated with activity in the left DLPFC, an area typically involved in higher order brain functions. Schlund, Hoehn-Saric, and Cataldo (2007) have commented that procedural and analytic issues may limit interpretations of Dickins et al.'s results. In brief, Dickins et al. employed a fixed, rather than a random, factors analysis to analyze their fMRI data and therefore their inferences apply only to the sample studied rather than the entire population. Furthermore, the findings of Dickins et al. were based on block analyses of 15-s blocks of time that contained multiple MTS trials. For example, in one block, transitivity trials (A-C relations) were tested and were compared with blocks in which the participant performed identity matching. This "block" MRI design was more powerful with respect to detecting regional variations in BOLD contrast, but included activation to irrelevant events (e.g., the inter-trial interval).

Schlund et al. (2007) trained 12 participants, using an MTS protocol, to establish two three-member equivalence relations. Subsequently, during fMRI at 3 T, participants were presented with trained relations and symmetrical, transitive and equivalent relations in an event-related design. MTS trials involving matching two identical circles served as the control condition. Schlund et al.'s analyses were event-related, focusing only on MTS trial onset, and were restricted a priori to anatomically defined areas (thus reducing the number of multiple comparisons). Results of second-level random effects analyses showed the left DLPFC was found to be active for all relations versus the control condition. Trained relations were associated with left ventrolateral prefrontal cortex (VLPFC) and left inferior parietal activation. In contrast, right hemisphere activation was observed in the inferior parietal lobule to directly trained, transitive, and equivalence relations. The DLPFC region was active to trained, symmetry, and transitive relations. The VLPFC region showed activation to trained, and transitive (but not symmetry or equivalence) relations. Thus, Schlund et al. demonstrated that potential dissociations may exist among different types of derived relations. Both DLPFC and inferior parietal lobule activation are consistent with results reported in numerous human imaging studies that involve the remembering of previously observed stimuli and stimulus relations, and manipulation of new information during TI (Wendelken, Bunge, & Carter, 2008).

Schlund, Cataldo, and Hoehn-Saric (2008) again employed a MTS protocol to train two three-member equivalence classes to examine medial temporal lobe involvement in DRR. During fMRI scanning at 3 T, participants were asked to judge if two presented stimuli were related. Testing included symmetrical, transitive and equivalent pairs. Cross-class "foils"—consisting of unrelated stimuli—were also presented as the control condition. Of the notable findings in this study, hippocampal activation was observed for transitivity and for equivalence (combined symmetry and transitivity) but not for symmetrical relations. Typically, such activation has been observed in TI studies in which stimuli are ordered along a dimension (e.g., John is smaller than Bill, Bill is smaller than Tom). Schlund et al. (2008) concluded that hippocampal activation is dependent on there being an intervening stimulus between stimuli, rather than a serial order involving a comparative relation *per se.*

Ogawa, Yamazaki, Ueno, Cheng, and Iriki (2009) trained 15 participants to establish five three-member equivalence classes. Following conditional discrimination training, participants were scanned at 4 T and presented with symmetrical, transitive and equivalence relations. In contrast to Schlund et al. (2008), who tested the derived relations prior to scanning, participants in Ogawa et al. were exposed to the derived relations for the first time during scanning. Using a block design, each test block contained five derived relation test trials, and also five directly trained trials, presented in alternation. These mixed test and directly trained blocks were compared with blocks consisting of 10 directly trained trials. For example, BA (symmetry) test blocks had AB (directly trained) trials as the comparison block and so on for all other relations. Test trials were always presented in the following block order: BA, CB, AC, and CA. In a departure from previous equivalence/fMRI studies, Ogawa et al. also presented sham feedback (i.e., regardless of whether their responses were correct, the subjects were invariably presented with a green "O" indicating "correct."). For BA symmetrical relations, the prefrontal cortex (PFC), intraparietal area (IPA), and medial frontal cortex (MFC) were activated bilaterally. For symmetrical CB relations, which always followed the AB block, only right PFC was activated. Ogawa et al. suggested that the bilateral IPA activation reflected the active, immediate manipulation of the learned relations to make inferences about the novel relations and that this manipulation brings about the inversion or reversal of the learned relations to derive symmetry (hence, IPA activation was not observed in the CB session because the manipulation had already been completed in the BA session, although this comparison was not directly tested). Both transitivity and equivalence sessions showed brain activation similar to that in the BA symmetry session. In addition, the left DLPFC and right anterior PFC were activated in the transitivity block, and posterior cingulate cortex and precuneus were activated in the equivalence block. The right PFC was active for both types of relation that involved symmetry (i.e., a reversed relation)—this is not surprising as the right PFC plays a key role in executive functions. The MFC, associated with conflict monitoring and resolution, was activated differentially by exposure to learned and emergent relations.

In addition to equivalence relations, More-than and Less-than relations have also been studied using fMRI. Hinton, Dymond, Von Hecker and Evans (2010) included 12 participants in a group trained with All-More relations and 12 participants in a group trained with All-Less relations. Hinton et al. employed a similar training approach to Reilly et al. (2005), establishing a 5-term linear series of items ordered along a quantitative dimension. This approach was advantageous in that it facilitated the use of mutually entailed and combinatorial relations not usually studied in the TI literature. Hinton et al.'s RT results replicated the finding that RT decreases linearly with increasing distance between members of the relational network. Following fMRI at 3 T, a similar linear trend was found in inferior frontal cortex, DLPFC, and parietal cortex bilaterally. That is, the greatest activation was for adjacent relations and the least activation was for two-step relations. There was greater activation in the left parietal cortex and longer RTs when inferring test trials with contextual cues different than training. Bilateral hippocampal activation was seen when comparing each relation type to baseline. Interestingly, no significant differences in left or right hippocampal activation were found between relational types (trained, mutually entailed, combinatorial one-step or two step), or when split into More-than or Less-than relations. Hinton et al. concluded that the hippocampus may play a more general role during relational reasoning rather than inference *per se*, and perhaps the role of the hippocampus is to maintain the relational network (cf. Schlund et al., 2008).

Overview of fMRI Findings on DRR

Figure 1 displays some candidate brain regions suggested by studies that may comprise the foundational neurocircuitry that supports various forms of derived responding in humans. While there are some divergent findings across studies and formal meta-analyses have yet to be performed, the published findings suggest there may be a core set of brain regions recruited during DRR. The figure highlights activation in dorsal, ventral and medial portions of the prefrontal cortex and posterior parietal regions including the inferior parietal lobule and precuneus, as well as subcortical regions that include the hippocampus and striatum. A rather consistent finding across studies is the involvement of dorsal regions of the prefrontal cortex and posterior parietal regions, both of which have been implicated in the manipulation of stimuli and executive functioning (i.e., regulation of cognitive and behavioral responses). Further, given that non-human animals do not show equivalence (see Dymond, Roche, & Barnes-Holmes, 2003), some researchers (e.g., Ogawa et al., 2009) have suggested that PFC and IPA may constitute a uniquely human fronto-parietal network.

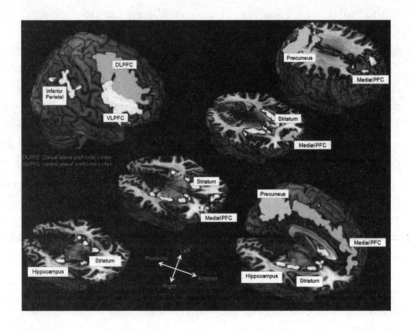

Figure 1. Candidate brain regions suggested by studies that may comprise the foundational neurocircuitry that supports various forms of derived responding in humans (see text for details). PFC = prefrontal cortex; DLPFC = dorsolateral PFC; VLPFC = ventrolateral PFC.

There were also some discrepant results in the fMRI studies described. For example, Schlund et al. (2008; equivalence relations) and Hinton et al. (2010; comparative relations) reported hippocampal

Table 1. Areas of Critical Procedural Difference across Imaging Studies on DRR

Study	Baseline	Stimuli	Magnet strength (T)	Analysis	Coverage	N	Second level analysis
Dickins et al. (2001)	Fluency task	Common icons	1.5	Block	Whole Brain	12	No
Schlund et al. (2007)	Identity Match	ASCII characters	3	Event Related	Region of interest	12	Yes
Schlund et al. (2008)	Cross Class Foils	Pseudowords	3	Event Related	Whole Brain	20	Yes
Ogawa et al. (2009)	Block Trained	Arbitrary pictures	4	Block	Whole Brain	15	Yes
Hinton et al. (2010)	Fixation	Pseudowords	3	Event Related	Whole Brain and Region of interest	12	Yes

involvement whereas Ogawa (2009: equivalence relations), Schlund et al. (2007: equivalence relations) and Dickins et al. (equivalence relations) did not report any differential hippocampal activity. Ogawa et al. argued that the interval between training and testing (3 hrs between training and scanning in Schlund et al., 2008, overnight in Ogawa et al.) could influence the brain areas responsible for retrieval and maintenance of the relations. There are, however, many potential reasons for different results across studies, not least due to the type of design or analysis, whether block or event-related. Table 1 outlines some procedural differences in the fMRI/DRR research to date, some of which are likely responsible for divergent patterns of brain activation across studies.

Implications for Cross-Disciplinary Studies

Measures of DRR such as RT, EEG and fMRI do not need to have any special status in an analysis, nor do they have to explain behavior "at some other level of observation " (Skinner, 1950, p. 193): rather, they are useful tools that can enhance the prediction and influence of target behaviors. These measures can also allow us to address topics of interest to other branches of psychology. For example, a key unanswered question concerns why nodal distance effects and symbolic distance effects occur in opposite directions: RT increases as nodal distance increases among stimuli related by coordination but decreases among stimuli related by comparison. A combined behavior-analytic/neuroscience approach may be uniquely positioned to address this issue. Traditional cognitive approaches typically use words from natural language categories. Mediated priming using natural language (e.g., STRIPES primes LION through the mediating stimulus TIGER) is difficult to demonstrate, at least in part because the word STRIPES is not uniquely related to TIGER. Similarly, training direction (e.g., training using a More-than relation then testing a Less-than relation) cannot be manipulated using pre-existing categories. In contrast, behavioral approaches employ intra-experimentally generated stimulus relations with arbitrary stimuli, and thus the history of behavior in question can be more precisely controlled.

Confounding variables such as syntax, polysemy and each subject's idio-syncratic pre-experimental semantic history can be minimized.

Mounting evidence suggests the fronto-parietal network could be a significant target for future DRR research that involves not only neuro-imaging but also clinical neuroscience approaches concerned with understanding the effects of focal lesions, such as those produced by strokes, or structural abnormalities that produce executive functioning deficits. Of particular importance for future neuroimaging studies is the need to investigate the *functional connectivity* between regions high-lighted in studies by exploiting information about the regional time course of the brain response during DRR. Identifying the critical neural pathways for successful DRR could have implications for our understand-ing of conditions in which these faculties are impaired. For example, there is a literature to support the notion that *functional underconnectiv-ity* is a fundamental feature of autism (Belmonte et al., 2004). Indeed, a frontal-posterior (temporoparietal junction) deficit in connectivity has already been observed in children with autism as they complete a *Theory of Mind* task (Kana, Keller, Cherkassky, Minshew, & Just, 2009). Given the finding that Theory of Mind abilities (i.e., deictic framing) can be trained (Weil, Hayes, & Capurro, 2011), it is possible that targeting deictic frames, and perhaps training relational framing flexibility in general, in early development, could aid the development of fronto-pos-terior connections.

To broaden the accessibility and acceptance of DRR procedures, we propose that demonstrating the validity and reliability of existing and novel DRR measures and methods is more likely to facilitate collabora-tions with researchers from other disciplines. For example, protocol anal-ysis, in which participants concurrently report their behavior, is a poten-tially useful measure that has been underutilized to date. Sorting tests are also underrepresented, and could easily be integrated into existing protocols, such as requiring participants to freely drag-and-drop pairs of stimuli into an area of the screen designated by a contextual cue (e.g., drag a pair of stimuli related by opposition into the area of the screen designated by the OPPOSITE contextual cue). DRR research may also benefit from application of confidence estimates of "relatedness". One alternative approach might involve recall paradigms where responding involves a form of reproduction (e.g., drawing or speaking). We also know little about DRR and relational strength when relations are acquired and transformed through socially mediated processes, such as instructions or observational learning. In any case, such efforts are likely to forge important ties with other disciplines by employing a common currency which functions as the foundation for meaningful collaboration.

Conclusions

The goal of this chapter was to highlight the growth and diversity of DRR research and bring to light some of the novel behavioral and neurophysiological methods and measures that have been developed and employed to give us new perspectives of DRR. We hope this provides behavioral researchers a compass for gaining a sense of the origins of DRR research and directions taken by various forms of DRR research, along with suggestions for the future. Initial studies of DRR often focused on the acquisition and boundary conditions of relational responding, typically equivalence, that could be addressed using MTS preparations and accuracy measures. This research base provided the necessary footing for testing increasingly sophisticated hypotheses, in areas involving multiple stimulus relations such as opposition and comparative relations, as well as the role of DRR in cognitive processes generally thought to mediate nodal distance, priming, and analogical responding. As additional dimensions of DRR were revealed through the refinement of novel training and testing methods, with formats such as the REP and RCP, it became possible to integrate DRR methods and measures with emerging neuroscience methods to address questions about brain-behavior relations and functional-anatomic correlates of DRR.

In summary, there appears to be many good reasons to be optimistic about the potential contributions of RFT and varieties of DRR research in behavioral and neurophysiological research. However, it seems important to remain vigilant and sensitive to increasing calls for high quality translational research with real face and ecological validity. But perhaps most importantly, we would argue that the longevity and utility of RFT and continued DRR research should not have as its primary goal to develop new questions and novel methods and measures of DRR. Rather, the behavioral research community should continue to commit to developing and applying new methods and measures to try to answer questions we share with other scientific disciplines. This strategy of "bridge building" seems a more worthy long-term strategic plan and one which avoids the "building a better mouse trap" fallacy.

References

Acuna B. D., Eliassen J. C., Donoghue J. P., & Sanes J. N. (2002). Frontal and parietal lobe activation during transitive inference in humans. *Cerebral Cortex, 12,* 1312–1321.

Arntzen, E. (2004). Probability of equivalence formation: Familiar stimuli and training sequence. *The Psychological Record, 54,* 275-291.

Barnes-Holmes, D., Staunton, C., Whelan, R., Barnes-Holmes, Y., Commins, S., Walsh, D., Stewart, I., Smeets, P. M., & Dymond, S. (2005). Derived stimulus relations, semantic priming, and event-related potentials: Testing a behavioral theory of semantic networks. *Journal of the Experimental Analysis of Behavior, 84,* 417-433.

Barnes-Holmes, D., Regan, D., Barnes-Holmes, Y., Walsh, D., Stewart, I., Smeets, P. M., Whelan, R., & Dymond, S. (2005). Relating derived relations as a model of analogical reasoning: Reaction times and event related potentials. *Journal of the Experimental Analysis of Behavior, 84,* 435-451.

Belmonte, M. K., Allen, G., Beckel-Mitchener, A., Boulanger, L. M., Carper, R. A., & Webb, S. J. (2004). Autism and abnormal development of brain connectivity. *Journal of Neuroscience, 24,* 9228–9231.

Cabello, F., & O'Hora, D. (2002). Addressing the limitations of protocol analysis in the study of complex human behavior. *International Journal of Psychology and Psychological Therapy, 2,* 115-130.

Collins, A. M., & Loftus, E. F. (1975). A spreading-activation theory of semantic processing. *Psychological Review, 82,* 407-428.

Cullinan, V., Barnes, D., Hampson, P. J., & Lyddy, F. (1994). A transfer of explicitly and non-explicitly trained sequence responses through equivalence relations: An experimental demonstration and connectionist model. *The Psychological Record, 44,* 559-585.

Cullinan, V., Barnes-Holmes, D., & Smeets, P. M. (2001). A precursor to the relational evaluation procedure: Searching for the contextual cues that control equivalence responding. *Journal of the Experimental Analysis of Behavior, 76,* 339–349.

Dickins, D. W., Singh K. D., Roberts N., Burns, P., Downes, J. J., Jimmieson, P., & Bentall, R. P. (2001). An fMRI study of stimulus equivalence. *Neuroreport, 12,* 405–411.

Dymond, S., & Rehfeldt, R. A. (2001). Supplemental measures of derived stimulus relations. *Experimental Analysis of Human Behavior Bulletin, 19,* 8-12.

Dymond, S., Roche, B., & Barnes-Holmes, D. (2003). The continuity strategy, human behavior, and behavior analysis. *The Psychological Record, 53,* 333-347.

Dymond, S., Roche, B., Forsyth, J. P., Whelan, R., & Rhoden, J. (2007). Transformation of avoidance response functions in accordance with the relational frames of same and opposite. *Journal of the Experimental Analysis of Behavior, 88,* 249–262.

Dymond, S., Roche, B., Forsyth, J. P., Whelan, R., & Rhoden, J. (2008). Derived avoidance learning: Transformation of avoidance response functions in accordance with the relational frames of same and opposite. *The Psychological Record, 58,* 271–288.

Dymond, S., & Whelan, R. (2010). Derived relational responding: A comparison of matching-to-sample and the relational completion procedure. *Journal of the Experimental Analysis of Behavior, 94,* 37–55.

Fields, L., & Verhave, T. (1987). The structure of equivalence classes. *Journal of the Experimental Analysis of Behavior, 48,* 317–332.

Fields, L., & Watanabe-Rose, M. (2008). Nodal structure and the partitioning of equivalence classes. *Journal of the Experimental Analysis of Behavior, 89,* 359–381.

Guinther, P. M., & Dougher, M. J. (2010). Semantic false memories in the form of derived relational intrusions following training. *Journal of the Experimental Analysis of Behavior, 93,* 329–347.

Haimson, B., Wilkinson, K. M., Rosenquist, C., Ouimet, C., & McIlvane W. J. (2009). Electrophysiological correlates of stimulus equivalence. *Journal of the Experimental Analysis of Behavior, 92,* 245–256.

Hayes, S. C., & Barnes, D. (1997). Analyzing derived stimulus relations requires more than the concept of stimulus class. *Journal of the Experimental Analysis of Behavior, 68,* 235–244.

Hayes, S. C., & Bissett, R. (1998). Derived stimulus relations produce mediated and episodic priming. *The Psychological Record, 48,* 617-630.

Hinton, E. C., Dymond, S., von Hecker, U., & Evans, C. J. (2010). Neural correlates of relational reasoning and the symbolic distance effect: Involvement of parietal cortex. *Neuroscience 168,* 138–148.

Horne, P. J., & Lowe, C. F. (1996). On the origins of naming and other symbolic behavior. *Journal of the Experimental Analysis of Behavior, 65,* 185–241.

Horne, P. J., & Lowe, C. F. (1997). Toward a theory of verbal behavior. *Journal of the Experimental Analysis of Behavior, 68,* 271–296.

Kana, R. K., Keller, T. A., Cherkassky, V. L., Minshew, N. J., & Just, M. A. (2009). Atypical frontal-posterior synchronization of theory of mind regions in autism during mental state attribution. *Social Neuroscience, 4,* 135–152.

Lipkens, R., & Hayes, S. C. (2009). Producing and recognizing analogical relations. *Journal of the Experimental Analysis of Behavior, 91,* 105–126.

Munnelly, A., Dymond, S., & Hinton, E. C. (2010). Relational reasoning with derived comparative relations: A novel model of transitive inference. *Behavioural Processes, 85,* 8–17.

Murphy, C., Barnes-Holmes, D., & Barnes-Holmes, Y. (2005). Derived manding in children with autism: Synthesizing Skinner's *Verbal Behavior* with relational frame theory. *Journal of Applied Behavior Analysis, 38,* 445-462.

Ogawa, A., Yamazaki, Y., Ueno, K., Cheng, K., & Iriki, A. (2009). Neural correlates of species-typical illogical cognitive bias in human inference. *Journal of Cognitive Neuroscience, 22,* 2120-2130.

O'Hora, D., Roche, B., Barnes-Holmes, D., Smeets, P. M. (2002). Response latencies to multiple derived stimulus relations: testing two predictions of relational frame theory. *The Psychological Record, 52,* 51–75.

O'Hora, D., Barnes-Holmes, D., Roche, B., & Smeets, P. M. (2004). Derived relational networks as novel instructions: A possible model of generative verbal control. *The Psychological Record, 54,* 437–460.

Posner, M. I. (2005). Timing the brain: Mental chronometry as a tool in neuroscience. *PLoS Biology, 3,* e51.

Reilly, T., Whelan, R., & Barnes-Holmes, D. (2005). The effect of training structure on the latency of responses to a five-term linear chain. *The Psychological Record, 55,* 233–249.

Roche, B., Linehan, C. Ward, T, Dymond, S., & Rehfeldt, R. A. (2004). The unfolding of the relational operant: A real-time analysis using electroencephalography and reaction time measures. *International Journal of Psychology and Psychological Therapy, 4,* 587-603.

Schlund, M. W., Hoehn-Saric, R., & Cataldo M. F. (2007). New knowledge derived from learned knowledge: Functional-anatomic correlates of stimulus equivalence. *Journal of the Experimental Analysis of Behavior, 87,* 287–307.

Schlund M. W., Cataldo, M. F., & Hoehn-Saric, R. (2008). Neural correlates of derived relational responding on tests of stimulus equivalence. *Behavioral and Brain Functions, 4,* 6.

Sidman, M. (1994). *Equivalence relations and behavior: A research story.* Boston, MA: Authors Cooperative.

Skinner, B. F. (1950). Are theories of learning necessary? *Psychological Review, 57,* 193–216.

Smeets, P. M., Dymond, S., & Barnes-Holmes, D. (2000). Instructions, stimulus equivalence, and stimulus sorting: Effects of sequential testing arrangements and a default option. *The Psychological Record, 50,* 339-354.

Stewart, I., Barnes-Holmes, D., & Roche, B., & Smeets, P. M. (2002). A functional analytic model of analogy. *Journal of the Experimental Analysis of Behavior, 78,* 375-396.

Stewart, I., Barnes-Holmes, D., & Roche, B. (2004). A functional-analytic model of analogy using the relational evaluation procedure. *The Psychological Record, 54,* 531–552.

Wang, T., Dack, C., McHugh, L., & Whelan, R. (2011). Preserved nodal number effects under equal reinforcement. *Learning and Behavior, 39,* 224–238.

Wang, T., McHugh, L., & Whelan, R. (2012). A test of the discrimination account in equivalence class formation. *Learning and Motivation, 43,* 8-13.

Weil, T. M., Hayes, S. C., & Capurro, P. (2011). Establishing a deictic relational repertoire in young children. *The Psychological Record, 61,* 371-390.

Wendelken, C., Bunge, S. A., & Carter, C. S. (2008). Maintaining structured information: An investigation into functions of parietal and lateral prefrontal cortices. *Neuropsychologia, 46,* 665–678.

Whelan, R., & Barnes-Holmes, D. (2004). The transformation of consequential functions in accordance with the relational frames of same and opposite. *Journal of the Experimental Analysis of Behavior, 82,* 177–195.

Whelan, R., Barnes-Holmes, D., & Dymond, S. (2006). The transformation of consequential functions in accordance with the relational frames of more-than and less-than. *Journal of the Experimental Analysis of Behavior, 86,* 317–335.

Whelan, R., Cullinan, V., O'Donovan, A., & Rodriguez Valverde, M. (2005). Derived same and opposite relations produce association and mediated priming. *International Journal of Psychology and Psychological Therapy, 5,* 247–264.

Whelan, R. (2008). Effective analysis of reaction time data. *The Psychological Record, 58,* 475–482.

Yorio, A., Tabullo, A., Wainselboim, A., Barttfeld, P., & Segura, E. (2008). Event-related potential correlates of perceptual and functional categories: Comparison between stimuli matching by identity and equivalence. *Neuroscience Letters, 443,* 113–118.

CHAPTER FIVE

A Functional Approach to the Study of Implicit Cognition: The Implicit Relational Assessment Procedure (IRAP) and the Relational Elaboration and Coherence (REC) Model

Sean Hughes

Dermot Barnes-Holmes

National University of Ireland, Maynooth

I n the following chapter we review the methodological, empirical and theoretical developments that are currently shaping the landscape of implicit cognition research. We begin by mapping out the methods and theories cognitive researchers have traditionally adopted and then examine how researchers within contextual behavioral science (CBS) have also sought to understand, predict and influence these same behaviors using a unique set of conceptual and procedural tools. In particular, we discuss how Relational Frame Theory has served to generate one methodology, termed the Implicit Relational Assessment Procedure (IRAP), and one possible functional account of implicit cognition, known as the Relational Elaboration and Coherence (REC) model. We will attempt to show that this work has fostered a rich and progressive account of implicit cognition. Finally, we close by considering several pressing empirical and conceptual issues that the IRAP and REC model face going forward as well as future directions for research in this area.

The Cognitive Approach to the Study of Implicit Cognition

Over the past 15 years the study of implicit cognition has occupied center stage in social psychology and flourished into a substantial and distinct research topic throughout psychological science (for an overview, see Banaji & Heiphetz, 2010; Nosek, Hawkins, Frazier, 2011; Payne & Gawronski, 2010). Broadly speaking, much of this work is rooted in the finding that people behave in two different and potentially conflicting ways. On the one hand, and consistent with our intuitive beliefs about behavior, we can respond to stimuli in our environment in a thoughtful, deliberate and controlled manner. These "explicit" evaluations are argued to be "controllable, intended, made with awareness, and [require] cognitive resources" (Nosek, 2007, p.65) and are typically registered through the use of direct procedures such as semantic differential scales, Likert scales, interviews, and focus groups. On the other hand, our history of interacting with the social environment can also result in the formation of evaluative responses that occur automatically and without either intention or control (termed "implicit evaluations"; see Gawronski & Payne, 2010)[1]. Automatic and deliberate evaluative responses often cohere with one another when phobic stimuli (Teachman, 2007), consumer preferences (Maison, Greenwald & Bruin, 2004) and political orientation (Choma & Hafer, 2009) are subject to inquiry. However, an impressive amount of data now indicates that under specific conditions—and with respect to certain stimuli—automatic and deliberate evaluative responding may conflict. For instance, people often demonstrate an automatic negative response to members of other racial, ethnic

---

1   Following the recommendations of De Houwer (2006), we define a measurement procedure as either *direct* or *indirect* on the basis of its procedural properties, and the outcome or effect of a procedure as either *implicit* or *explicit* on the basis of the assumed properties of the psychological attribute being measured.

or religious groups despite their self-reported egalitarian sentiments (e.g., Greenwald, Poehlman, Uhlman & Banaji, 2009; Payne, Burkley, & Stokes, 2008).

In light of the fact that automatic and deliberate evaluative responses can converge or diverge as well as differentially predict important real-world behaviors (see Perugini, Richetin, & Zogmaister, 2010, for a detailed overview), research has increasingly focused on how these two types of responding are established, manipulated, and changed. Addressing these various questions has served to reshape the methodological and theoretical landscape of social psychological research. For instance, the need to capture automatic behaviors uncorrupted by self-presentation biases has led to the emergence of a new heterogeneous class of indirect procedures that includes semantic and evaluative priming (Fazio, Jackson, Dunton, & Williams, 1995; Wittenbrink, Judd, & Park, 1997), the Go/No-go Association Test (gNAT; Nosek & Banaji, 2001), the Affect Misattribution Procedure (AMP; Payne, Cheng, Govorun, & Stewart, 2005) and the Implicit Association Test (IAT; Greenwald, McGhee, & Schwartz, 1998) amongst others. Although methodologically diverse, indirect procedures generally aim to provide an estimate of automatic (evaluative) responding through either the speeded categorization of stimuli or subjective judgments of ambiguous stimuli.

These methodological innovations have not only led to a wealth of empirical data but have also unleashed a new wave of theorizing about *when* and *why* performance on direct and indirect procedures will converge, diverge or predict behavior. More often than not, these explanations have been guided by a mediational, mechanistic approach to psychological science (i.e., operate at the cognitive level of analysis). A core philosophical assumption underpinning this approach is the notion that behavior is causally mediated through the action and interplay of certain mental process and representations. The researcher's analytic goal involves identifying and manipulating mental constructs in order to predict behavior and its change. Importantly, however, mental processes/representations can be neither contacted nor manipulated in space or time but only inferred through relevant changes in behavior (see De Houwer, 2011). Consequently, this meta-theoretical strategy of treating behavior as a proxy for mental events has given rise to a large number of competing cognitive theories that differ in the nature and number of mental processes they postulate and how those processes interact with and account for behavior. For example, the behavioral outcomes of direct and indirect procedures have been qualified in many different ways, from the operation of a single type of mental representation (e.g., object-evaluation association in memory) acting under different conditions (Olson & Fazio, 2009; Petty, Briñol & De Marree, 2007) to multiple interacting memory systems (e.g., Amodio & Ratner, 2011). That said, dual-process models have traditionally dominated cognitive theorizing in this research area (see Gawronski & Creighton, in press), with behavior obtained from direct and indirect procedures broadly classified as the product of two qualitatively distinct—yet potentially interacting—kinds of mental constructs called associations and propositions (e.g., Gawronski & Bodenhausen, 2011; Rydell & McConnell, 2006; Strack & Deutsch, 2004; see Hughes, Barnes-Holmes & De Houwer, 2011, for a detailed discussion).

In short, the emergence of indirect procedures has constituted nothing short of a modern measurement revolution and transformed the study of human social cognition. This new industry of methodologies has sparked an accelerated accumulation of evidence and triggered a new wave of meditational, mechanistic theorizing. As noted above, the primary analytic goal of the cognitive approach is to develop, test and refine theories that explain how mental processes and representations are formed, activated and changed as well as influence subsequent behavior. Without doubt this conceptual framework has afforded considerable utility in helping us to understand automatic and deliberate responding—yet it is not without its limitations. In particular, it has been argued that the practice of treating the behavioral outcomes generated by measurement procedures as proxies for mental constructs (e.g., associations/propositions) may serve to constrain theoretical and methodological diversity as well as scientific progress more generally (e.g., De Houwer, 2011; Hughes et al., 2011). At the same time, it is important to note that other philosophical frameworks may also yield legitimate and fruitful lines of inquiry. As the current volume attests, functional contextualism represents one such philosophy of science that is committed to understanding human behavior using a distinct set of conceptual and methodological tools.

A Functional Approach to the Study of Implicit Cognition

Broadly speaking, functional contextualism (referred to hereafter as the "functional approach") is a pragmatic philosophy of science with one unified goal: to understand, predict, and influence behavior with scope (explain a comprehensive range of behaviors across a variety of situations), precision (applying a restricted set of principles to any event) and depth (cohere across analytical levels and domains such as biology, psychology, and anthropology). What differentiates this approach from others such as cognitivism is that it adopts an exclusively functional epistemology. Rather than locating the causation of behavior in a mental domain, scientific analysis is focused on the functional relations between the (past and present) environment and behavior that unfold across both time and context. In other words, functional researchers treat behavior as an "act-in-context" and this context can be private or public and vary from the most immediate behavioral instance to temporally delayed and historical behavioral sequences (Biglan & Hayes, 1996; Gifford & Hayes, 1999).

One functional theory that sets out to explain a wide variety of public and private behaviors using only a handful of interrelated principles is Relational Frame Theory (RFT; Hayes, Barnes-Holmes & Roche, 2001), the central topic of the current volume. In the sections that follow we consider how RFT has facilitated a rich and progressive functional approach to the study of automatic cognition through the development of the IRAP and REC model. Before proceeding, we would like to clarify one issue. Whereas previous IRAP research has often used the terms "automatic", "implicit", and "brief and immediate" interchangeably, both "automatic" and "implicit" are typically defined in terms of the presence and/or operation of specific cognitive processes (e.g., awareness, intention or efficiency; Bargh, 1994). Therefore, given that (a) the functional account of "implicit" and "automatic" precludes mediating mental mechanisms coupled with (b) the fact that a functional definition of automatic/implicit has yet to be fully articulated (although see De Houwer & Moors, 2012), we will use "brief and immediate" to refer to this particular class of behaviors throughout the rest of the current chapter.

Relational Elaboration and Coherence (REC) Model

The REC model draws upon a core theoretical premise of RFT—that human language and cognition are relational in nature—to account for the behaviors captured by direct and indirect procedures (Barnes-Holmes, Barnes-Holmes, Stewart, & Boles, 2010). According to this account, relational responses, like all behaviors, unfold across time. Thus, when a stimulus is encountered, a relational response may occur relatively quickly and be followed by additional relational responses. These additional relational responses may occur toward the stimulus itself or toward the initial response to that stimulus. With sufficient time, these additional relational responses will likely form a coherent relational network (discussed below).

For illustrative purposes, imagine that a person is on an airplane and has to quickly decide where to sit before take-off. However, only two seats remain available—in one seat is a member of the person's own ethnic group (e.g., Irish male) and in another a member of an out-group (e.g., Arabic male). The first relational response to occur for this person might involve a negative evaluation based on a verbal history in which Arabic men are portrayed (in the media) as dangerous. However, additional relational responding may involve a different evaluation, such as "He may not be dangerous" and/or "reacting on the basis of ethnicity is wrong." In this instance, the person's subsequent behavior (i.e., where he or she sits) may be a consequence of their brief and immediate relational responses (BIRRs) or the more extended and elaborated relational responses (EERRs) that typically occur thereafter. Although the foregoing example suggests that BIRRs generally precede EERRs, it is important to appreciate that both are behavioral patterns, and thus they may interact in a dynamic manner. For example, an EERR may well generate a BIRR in a given stream of verbal behavior. Imagine that the traveler in the above example emits the EERR "It is wrong to discriminate on the basis of ethnicity," which then results in a BIRR involving a brief negative self-evaluative response (e.g., "I am bad").

From a REC perspective, it is these BIRRs and EERRs, broadly speaking, that form the basis of what cognitive researchers commonly label implicit and explicit evaluations. Consequently, and consistent with a functional epistemology, the behaviors captured by direct and indirect procedures are not viewed as the product of mediating mental constructs. Rather they are argued to reflect the operation of the same behavioral process (i.e., arbitrarily applicable relational responding) operating under two broadly different conditions (e.g., presence versus absence of time pressure to respond). Indeed, it is important to note that in making a distinction between BIRRs and EERRs we are not claiming that they constitute different or separate behavioral processes. Rather, we are parsing out the single process of relational framing in terms of a temporal parameter. In very general terms, evidence indicates that the term BIRR applies to the first few seconds of responding to a stimulating event, whereas the term EERR typically applies to responses that occur across longer periods of time (e.g., Barnes-Holmes, Murphy, Barnes-Holmes, & Stewart, 2010).

The REC model also offers an explanation for why performance on direct and indirect procedures will either converge or diverge by appealing to a property of relational responding termed relational coherence. Specifically, a relational network (i.e., a set of inter-related stimulus relations) is said to "cohere" when all of the individual elements relate to each other in a manner that is consistent with the reinforcement history typically provided by the verbal community for such relational responding. According to RFT, the verbal community constantly reinforces coherence (and punishes incoherence) within relational networks, to the extent that relational coherence itself becomes a type of conditioned reinforcer for most language users. Imagine, for example, that you read the statement, "A is more than B and B is more than C, but C is more than A." In this case, it is likely that you would recognize the incoherent nature of this simple relational network and question its veracity (Hayes et al., 2001).

This search for relational coherence applies equally to BIRRs and EERRs. Taking the previous example, responding to an Arabic man on an airplane as "dangerous" may not cohere with other, subsequent relational responses that follow this initial response (e.g., "I treat everyone equally regardless of ethnicity"). In this instance, brief and immediate versus extended and elaborated responding will conflict with one another, and additional relational activity may be needed to reduce or resolve this incoherence. For example, this may involve the person responding to his or her initial relational response as "wrong", and thus a divergence between responding on direct and indirect procedures should be observed. In other words, the REC model argues that people may "reject" BIRRs if they do not cohere with elaborate and extended relational responding. Nevertheless, in many instances BIRRs and EERRs will be similar and thus performance on direct and indirect tasks should converge. In the example above, the individual may conclude that the Arabic man does, in fact, look rather dangerous, which would thus cohere with the initial relational response and potentially govern what seat he or she chooses.

In short, the REC model explains not only the behaviors captured by direct and indirect procedures but also their convergence and divergence by appealing to the degree of coherence between relational responses and their relative speed. Broadly speaking, and in line with the pragmatic goal of the functional approach, relational responding can be characterized as either brief and immediate or extended and elaborated. Brief and immediate relational responses may or may not cohere with subsequent relational responding. When they cohere, direct and indirect procedures will typically converge, but when they do not, such procedures will typically diverge.

The Implicit Relational Assessment Procedure

Methodology

When implicit cognition is defined functionally (in terms of relational responding), a non-associative indirect procedure is not only possible but necessary. One such methodology capable of targeting BIRRs is the Implicit Relational Assessment Procedure (IRAP; Barnes-Holmes et al., 2006). In essence, the conceptual and methodological rationale for the IRAP was rooted in the desire to investigate the role of stimulus equivalence and derived relational responding in naturally occurring verbal relations. In order to make these verbal histories apparent within the laboratory, researchers strategically designed procedures to put pre-existing verbal relations into competition with laboratory-induced equivalence classes (see Watt, Keenan, Barnes, & Cairns, 1991, for a seminal study in this area). Over the past two decades this analytic strategy of juxtaposing natural with laboratory established verbal relations has yielded a fruitful functional analysis of a range of different psychological phenomena, including clinical anxiety (Leslie et al., 1993), social discrimination (Dixon, Rehfeldt, Zlomke, & Robinson, 2006) and human intelligence (O'Toole & Barnes-Holmes, 2009). Similarly, it has been used in the study of self-knowledge (Merwin & Wilson, 2005) and for identifying histories of child sexual abuse (McGlinchey, Keenan, & Dillenburger, 2000), as well as discriminating sex offenders from non-sex offenders (Roche, Ruiz, O'Riordan, & Hand, 2005). In a related line of inquiry, behavioral researchers have also attempted to develop behavior-analytic models of the IAT, or variations thereof (e.g., Barnes-Holmes et al. 2004; Gavin, Roche & Ruiz, 2008; O'Toole, Barnes-Holmes, & Smyth, 2007; O'Reilly, Roche, Ruiz, Tyndall & Gavin, 2012).

Arguably one of the most empirically and theoretically productive applications of the behavior-analytic approach to implicit cognition has been in the development of the IRAP. Simply put, the IRAP was designed to target a history of verbally relating specific classes of stimuli. The task involves contriving an experimental situation whereby an individual's pre-experimentally established verbal relations are pitted against those that are deemed inconsistent with that history of responding. This is achieved by presenting stimuli together with relational terms such as "True" and "False" and requiring participants to respond both quickly and accurately. In asking participants to respond to the relationship between two stimuli, each IRAP trial will likely cause the participant to emit a brief and immediate relational response prior to pressing the appropriate computer key. The probability of this response will be determined by the participant's prior learning history combined with current contextual variables. By definition, the most probable response will be emitted first most often. Therefore, on IRAP trials correctly designated as consistent with the participant's learning history and current contextual variables, the required key press will coordinate with the emitted response—producing faster response latencies and more accurate responses. Alternatively, inconsistent IRAP trials require a key press that opposes the first relational response emitted by the individual. Resolving this incoherence between covert relational responding and the required overt response takes time. Therefore, across multiple trials, the average latency for inconsistent blocks will be longer than for consistent blocks.

Put simply, the IRAP operates on the assumption that participants should be faster and more accurate when they have to relate stimuli in a manner deemed consistent (relative to inconsistent) with their prior learning history because brief and immediate relational responding will coordinate more often with consistent overt responding. The difference in time taken to respond on consistent relative to inconsistent trials— defined as the IRAP effect—is assumed to provide an index of the strength or probability of the targeted relations.

To illustrate this more clearly, consider the work of Roddy, Stewart and Barnes-Holmes (2011), who employed the IRAP to index brief and immediate verbal biases toward people's weight. On each trial, one of two label stimuli ("Good" or "Bad") was presented at the top of the computer screen with a target stimulus in the center of the screen (a photo of either an average-weight or an overweight person). Participants were required to choose between two response options ("Similar" and "Opposite") presented at the bottom left and right of the screen by pressing one of two keys. During a block of consistent trials, participants had to respond in a manner assumed to reflect prevailing verbal contingencies toward average and overweight people. For instance, choosing "Similar" given "Slim Person" and "Good" cleared the screen for 400 ms and presented the next trial. If an inconsistent response was emitted (e.g., choosing "Opposite" given "Slim Person" and "Good"), a red "X" appeared immediately under the target stimulus. To remove the red "X" and continue to the 400-ms inter-trial interval, participants were required to emit the consistent response. In contrast, for a block of inconsistent trials, participants are required to make an inconsistent response in order to progress from one trial to the next (e.g., choosing "Similar" given "Slim Person" and "Bad"). The IRAP typically consists of a minimum of two practice and a fixed set of six test blocks. In each case, every target stimulus is presented at least once in the presence of the two labels (with a minimum of six and a maximum of 12 target stimuli per label, allowing for a range of 24 to 48 trials per block). Although the same response contingency applies to all the trials within a particular block, these contingencies are reversed across successive blocks, with participants exposed to an alternating sequence of consistent and inconsistent responding. Participants are explicitly made aware of this fact in that once a block of trials is completed the program informs them that the relational contingencies in the next block will be reversed (thus removing any requirement for trial-and-error learning following the first block of trials).

To make the participant's history of brief and immediate relational responding apparent to the researcher, the IRAP requires the participant to respond with both speed (e.g., median response time of less than 2 s) and accuracy (at least 80% correct responses). These criteria also ensure that spurious contextual influences over relational responding are excluded as much as possible and task performance is primarily governed by the individual's prior verbal history (and relevant current contextual variables). Therefore, the practice phase of the IRAP allows participants up to four attempts to achieve these two criteria (i.e., a total of eight practice blocks) and if they fail to do so, they are thanked and debriefed and their data is discarded. Those participants who achieve the practice criteria proceed on to the six test blocks. Once again, failure to maintain speed and accuracy criteria across each of the test blocks results in the participant's data being excluded from subsequent analysis.

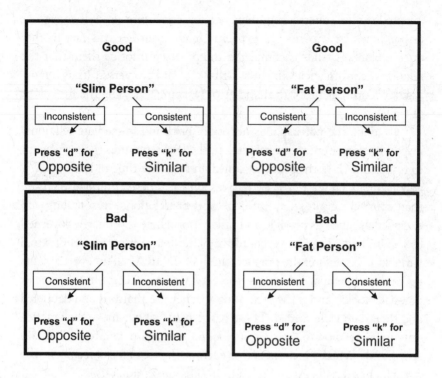

Figure 1. Examples of the four trial-types used in the Roddy et al. (2011) IRAP. The label stimuli ("Good" and "Bad"), target stimuli (pictures of a "Slim Person" and a "Fat Person") and relational response options (Similar and Opposite) are indicated in addition to the correct responses for each trial-type. The arrows and correct response option do not appear on screen or in the instructions.

One notable feature of the IRAP is that by presenting specific label and target stimuli together on a certain trial and requiring a particular response to be emitted, the IRAP can target four separate stimulus relations independent of one another. In the case of the Roddy et al. (2011) study discussed above, these stimulus relations (often called IRAP trial-types) were *Good-Slim*, *Good-Fat*, *Bad-Slim* and *Bad-Fat*. Therefore, on consistent blocks of IRAP trials participants were required to respond in a "pro-slim/anti-fat" manner and in precisely the opposite way for inconsistent blocks (i.e., "pro-fat/anti-slim"; see Figure 1 for an overview). Given this procedural design, the IRAP can provide two different measures of BIRRs depending on the researcher's analytic goal. On the one hand, by calculating the overall difference in response latency between consistent and inconsistent blocks, a global measure of BIRRing can be ascertained. Roddy et al., for instance, found an overall IRAP effect suggesting that participants responded quicker and more accurately to average-weight people as positive (and not negative) and overweight people as negative (and not positive). On the other hand, the IRAP can also provide a separate effect for each of the four different trial-types that make up the task. In doing so, the relationship between the four targeted stimulus relations can be independently determined. In the above example, a significant IRAP effect was observed for the *Good-Slim* and *Bad-Slim* trial-types, but there was no effect for the *Bad-Fat* and *Good-Fat* trial-types, indicating a positive verbal bias toward average-weight people but the absence of a negative bias toward their overweight counterparts (see Barnes-Holmes, Barnes-Holmes, Stewart & Boles, 2010, for a detailed overview of the procedure).

IRAP Research

Since its inception, the IRAP has stimulated a rapidly growing body of empirical work characterized by three key findings. First, and similar to other indirect procedures, the IRAP is capable of capturing behaviors that are not typically accounted for by self-report methodologies. In particular, performance on the IRAP has been found to diverge from, or appears inconsistent with, self-report data when participants are required to confirm verbal biases of a "psychologically sensitive" nature. For example, Dawson, Barnes-Holmes, Gresswell, Hart and Gore (2009) employed the IRAP in an attempt to assess BIRRs toward children as sexual in a sample of convicted sex offenders. Two label stimuli were used ("Adult" and "Child") in conjunction with a set of sexual or non-sexual target stimuli and two response options ("True" and "False"). Although no significant differences were obtained in self-reported beliefs between sex offenders and a control group of university students, the IRAP differentiated the two groups from one another in terms of their brief and immediate relational responding. Specifically, while the non-offender group demonstrated a strong response bias toward children as not sexual, the sexual offender group failed to make this distinction.

This divergence between brief and immediate relational responding on the one hand and self-reported responding on the other has now been replicated in a number of different attitude domains. Consider evaluations of members from other racial or ethnic groups. Across a number of IRAP studies, white participants professed egalitarian sentiments on a standard questionnaire of racial prejudice but also demonstrated a brief and immediate verbal bias favoring white people relative to their black counterparts (Barnes-Holmes, Murphy, et al., 2010; Power, Barnes-Holmes, Barnes-Holmes, & Stewart, 2009). Similarly, a discrepancy between self-reported preferences and the IRAP also emerges when religious stereotyping (Drake et al., 2010), social stereotyping (Cullen, Barnes-Holmes, Barnes-Holmes, & Stewart, 2009) and attitudes toward people's weight are subject to inquiry (Roddy, Stewart, & Barnes-Holmes, 2010; Roddy et al., 2011).

It should be noted, however, that IRAP and self-report responding typically converge when participants are not motivated to self-present or modify their behavior to concord with experimental or social expectations (i.e., when "non-socially sensitive" responding is targeted). Indeed, when attitudes toward country versus city living (Barnes-Holmes, Waldron, Barnes-Holmes, & Stewart, 2009), attitudes toward meat and vegetables in omnivores and vegetarians (Barnes-Holmes, Murtagh, Barnes-Holmes, & Stewart, 2010), self-esteem (Vahey, Barnes-Holmes, Barnes-Holmes, & Stewart, 2009; Timko, England, Herbert & Forman, 2010) and attitudes toward sexual stimuli are assessed (Stockwell, Walker & Eshleman, 2010), the IRAP demonstrates excellent convergent validity with self-reported responding. For instance, when the IRAP was used to examine undergraduates' self-esteem in the areas of intelligence, physical appearance, and friendliness/shyness, different IRAP trial-types correlated with the self-report data (Timko et al. 2010). In particular, faster and more accurate responding on the "I am [positive word]" trial-type was associated with better outcomes on explicit measures of quality of life, health and psychological/spiritual functioning, and lower outcomes on measures of depression, interpersonal sensitivity, and psychopathology.

A second key finding is that the IRAP demonstrates comparative, and in some instances, better levels of predictive validity to other established measures of BIRRs such as the IAT (Barnes-Holmes, Murtagh et al., 2010; Barnes-Holmes, Waldron et al., 2009), drug Stroop (Carpenter, Martinez, Vadhan, Barnes-Holmes, & Nunes, 2012), and EMG (Roddy et al., 2011). Furthermore, as noted above, the IRAP can target each of four stimulus relations under investigation (a conclusion that cannot be rendered from the relativistic nature of the IAT). To illustrate, consider a study by Timmins, Barnes-Holmes and Cullen (2011), in which the IRAP was used to discriminate between male respondents who self-identified as heterosexual versus homosexual (thus allowing a comparison to be made with previous work that used the IAT and a priming task; Snowden, Wichter & Gray, 2008).

Specifically, the Timmins et al. IRAP involved presenting a set of target words related to sexually attractive (e.g., "arousing", "erotic", "attractive") or sexually unattractive (e.g., "repulsive", "repelling", "repugnant") in the presence of nude images (both male and female) and the response options "True" and "False". Not only did performance on the IRAP predict participants' sexual orientation at levels comparable to the IAT and a priming task, but the non-relative nature of the procedure also permitted separate analyses of participants' brief and immediate responding to the male and female pictures. That is, heterosexual respondents demonstrated a significant attraction response bias for the female and repulsion bias for the male pictures, whereas the homosexual group showed the opposite pattern of responding for male stimuli. Likewise, when brief and immediate responding toward body size was examined in participants with eating disorders and a control group, the IRAP provided a detailed analysis of the independent relationships between the four stimulus relations (Parling, Cernvall, Stewart, Barnes-Holmes & Ghaderi, 2012). While both groups displayed pro-thin attitudes toward the self, the clinical group also showed a significantly stronger anti-fat attitude relative to the control group. In contrast, previous work using the IAT found pro-thin/anti-fat attitudes for both groups and no differences between groups (e.g., Schwartz, Vartanian, Nosek, & Brownell, 2006; Vartanian, Herman, & Polivy, 2005). Consequently, the relational rather than associative nature of the IRAP may provide important analytic advantages compared to relativistic alternatives such as the IAT.

Finally, not only does the IRAP capture BIRRs but it appears to be equal to or better than direct (or alternative indirect procedures) in predicting real-life behavior. Consider the work of Carpenter and colleagues

(2012), who administered a Drug Stroop protocol, the IRAP and a series of self-report questionnaires to a group of cocaine dependent participants before a 6-month outpatient treatment program. This work sought to determine the relationship between attentional biases toward cocaine stimuli, beliefs about the positive and negative effects of cocaine use and subsequent treatment outcome. Carpenter et al. found that participants' IRAP performance before the treatment program predicted their likelihood of attending as well as adhering to the program during the first 12 weeks of treatment. The Drug Stroop produced similar predictive validity for treatment outcomes during the second phase of treatment. In contrast, self-reported beliefs about or craving for cocaine did not predict either of these two treatment outcomes or correlate with IRAP or Stroop performance. In a similar vein, the IRAP has also been found to predict clinically relevant outcomes with respect to aversive responding in spider fearful individuals. Nicholson and Barnes-Holmes (2012b) set out to determine if the IRAP could distinguish between high and low spider fearful individuals and predict if they would approach a live tarantula. Consistent with past work using the IAT (e.g., Teachman, Gregg, & Woody, 2001) the IRAP successfully distinguished between the two groups such that high spider-fear participants demonstrated significantly greater aversive bias than their low-fear counterparts. Additionally, participants' BIRRs toward spiders on the IRAP predicted how likely they were to subsequently approach and interact with a live spider.

In sum, accumulating evidence indicates that the IRAP is characterized by good convergent validity given that it predicts certain classes of behavior similar to self-report methodologies. Furthermore, the IRAP also displays good divergent validity given its ability to predict behavior not captured by self-report methods, at levels similar to other behavioral (e.g., the IAT, priming, Drug Stroop) and physiological measures of BIRRing (e.g., EMG).

Refining the IRAP: An Ongoing Process

It is important to note that the level of experimental control achieved by the IRAP over brief and immediate relational responding is the product of seven years of continual refinement. Indeed, the ongoing modification of the original procedure (first reported in Barnes-Holmes et al., 2006), in response to empirical and theoretical developments, resulted in the emergence of what could be considered a "second-generation" IRAP. This iteration of the task is characterized by several features that distinguish it from its early predecessor. First, reducing the speed criteria from 3 s (as adopted in early IRAP research) to 2 s or less has served to increase not only the magnitude but more importantly the predictive validity of the task. Barnes-Holmes, Murphy et al. (2010) found that a significant racial bias emerged only when a 2-second rather than 3-second latency criterion was used. Furthermore, the internal reliability of their IRAP effect almost doubled in the 2-second condition, and results converged with findings from other indirect measures of racial bias (e.g., Greenwald et al., 2002). As such, increasing time pressure may serve to minimize the potential for contamination by extraneous verbal behavior over the brief and immediate relational responses that precede each overt IRAP response.

Second, in order to ensure that participants respond as quickly as possible and maintain that speed of responding throughout the task, trial-by-trial corrective feedback has also been introduced. In other words, the IRAP presents a warning message ("Too Slow") on any trial in which a participant fails to respond within the latency criterion. Although the impact of this second modification on both IRAP effect sizes and internal reliability has yet to be empirically determined, it appears anecdotally to increase both. Finally, the original IRAP simply used participants' response latencies in calculating the IRAP effect. However, more recent research suggests that this practice may be problematic for two reasons. On the one hand, response latency differences between consistent and inconsistent trials correlates highly with various measures of intelligence (O'Toole & Barnes-Holmes, 2009). On the other hand, demographic variables independent of the relational history being targeted (e.g., age, motor skills, cognitive ability) often contribute to longer average latencies and thus larger IRAP effects. In order to circumvent these problems, IRAP research has increasingly employed a modified version of the "D scoring algorithm" proposed by Greenwald, Nosek, and Banaji (2003). Using a D algorithm also facilitates a comparison between the IRAP and the IAT or other indirect procedures that also use this data transformation technique (see Barnes-Holmes, Barnes-Holmes, et al. 2010 for detailed discussion on the D-IRAP algorithm).

On balance, it is important to note that the data-analytic strategies described in this chapter reflect the practices that have evolved within a single research group, and as such should not be seen as prescriptive or necessarily the "best way" to conduct IRAP research or analyze subsequent data. In other words, the IRAP as a procedure and effect should not be conflated with any particular scoring algorithm. Indeed, the best way to analyze IRAP data should always remain largely an empirical issue (see Whelan, 2008).

In addition to these refinements, several other methodological issues have been explored. McKenna and colleagues (2007) provided compelling evidence that participants cannot "fake" the magnitude or direction of the IRAP effect even when they were given precise instructions on how to do so. Equally, the versatility of the procedure is evident in that it may be used to provide an index of verbal relations engineered within an experimental context (i.e., "attitude formation"; see Hughes & Barnes-Holmes, 2011) as well as test the relationship between relational responding and human intelligence (O'Toole & Barnes-Holmes, 2009). With respect to the latter study, the authors found that individuals who performed better on a standard measure of intelligence not only were faster at responding relationally on the IRAP but also demonstrated a greater degree of "relational flexibility" (i.e., alternating between consistent and inconsistent response contingencies). One implication extending from this work is that building flexible relational repertoires may play a crucial role in promoting intelligent behaviors (see also Cassidy, Roche & Hayes, 2011).

Challenges and Future Directions

Even though the IRAP's ability to target BIRRs has received strong empirical support in a number of different research areas, several procedural and conceptual issues need to be addressed. First, future work should continue to benchmark the validity and reliability of the task against well-established alternatives such as the IAT, AMP and affective and semantic priming. One means of accelerating this process is to ensure that all future IRAP studies report (a) accuracy and latency criteria, (b) number of participants who passed the IRAP, (c) correlations between indirect-indirect and/or indirect-direct procedures and (d) reliability (e.g., split-half reliability; test re-test reliability) and validity (e.g., predictive and construct where applicable). At present, this practice is unsystematic and largely varies from study to study (for an overview of the IRAP literature to date and the relevant procedural properties manipulated, see Table 1). Demonstrating that the IRAP targets and predicts the same type of behaviors as other indirect methodologies will also add weight to the theoretical argument that "automatic" cognition is relational (propositional) rather than primarily associative in nature (see Hughes, et al., 2011).

Second, a closer inspection is needed of the task-specific properties of the IRAP that shape the behavioral effects it generates. This is an important issue given that different indirect procedures have been found to show opposite or alternative outcomes for the same experimental manipulation (Deutsch & Gawronski, 2009). Thus, questions regarding fixed versus moving response options; the optimal number of stimuli to use in target and label classes; the optimal number of IRAP blocks; presence versus absence of corrective feedback and the impact of prior experience will need to be systematically addressed (see Campbell, Barnes-Holmes, Barnes-Holmes & Stewart, 2011 for initial work in this vein). Doing so will identify method-related moderators of the magnitude and predictive validity of IRAP effects.

Third, it has become increasingly apparent that the specific classes of stimuli inserted into the IRAP are critical in determining the internal reliability/ predictive validity of the measure. Ongoing research in our laboratory suggests that relatively minor differences in the target and label stimuli have the potential to produce subtle changes in IRAP performances. In a recent study, for example, participants completed two IRAPs with the same pictures of disgusting (e.g., dirty toilets) and pleasant (e.g., flowers) target stimuli but different label stimuli (Nicholson & Barnes-Holmes, 2012a). The label stimuli in one IRAP were designed to target *disgust propensity* (e.g., "I am disgusted") whereas the labels in the other IRAP targeted *disgust sensitivity* ("I need to look away"). Critically, both the disgust sensitivity and disgust propensity IRAPs predicted OCD behaviors and differentially predicted obsessing and washing concerns, respectively. Given the IRAP's apparent flexibility and precision in isolating specific effects in this way, IRAP researchers should remain fully alert to the importance of the stimuli that are used in their work.

Table 1: Overview of All IRAP Studies Conducted to Date

Note: N = Total number of participants in study; Topic = Research Area; Procedure = Procedure(s) used; N-IRAP = Number of IRAPs used in the study; N Dir = Number of Direct procedures used in the study; Speed = IRAP latency criteria; Acc = IRAP accuracy criteria; Pass Rate = Percentage of participants who passed the IRAP; D-IRAP = D-IRAP algorithm was used to analyze data; I.Con. = Internal consistency assessed; I-DC = Indirect-Direct Procedure Correlations; I-IC = Indirect-Indirect Procedure Correlations; P.V. = Predictive validity assessed. ? = Information not provided in paper.

Citation	N	Topic	Procedure	N-IRAP	N Dir	Speed	Acc	Pass Rate	D-IRAP	I.Con.	I-D C	I-I C	P.V.
Barnes-Holmes et al. (2006)	83	Attitudes	IRAP/EEG	3	3	3000	?	?					
McKenna et al. (2007)	36	Faking	IRAP	2	2	3000	80	?					
Barnes-Holmes et al. (2008)	40	Attitudes	IRAP/EEG	2	0	3000	80						
Barnes-Holmes et al. (2009)	26	Attitudes	IRAP/IAT	1	4	3000	80	80	X	X	X	X	
Chan et al. (2009)	131	Occupation	IRAP/IAT	1	2	3000	80		X	X	X	X	
Cullen et al. (2009)	24	Ageism	IRAP	1	2	3000	80	71			X		
Dawson et al. (2009)	32	Sexual	IRAP	1	1	5000	80	?	X		X	X	X
O'Toole & Barnes-Holmes (2009)	62	IQ	IRAP	2	1	N/A	80	89			X		X
Power et al. (2009)	32	National	IRAP	2	2	3000	80	?			X		
Vahey et al. (2009)	51	Self-Esteem	IRAP	1	1	3000	70	84	X		X		

Study	N	Topic	Method												
Barnes-Holmes, Murphy et al. (2010)	62	Race	IRAP	2	3	2000	80	81	X			X	X	X	X
Barnes-Holmes, Murtagh et al. (2010)	32	Food	IRAP/IAT	1	1	3000	80	?	X	X		X	X	X	X
Drake et al. (2010)	67	Attitudes	O-M IRAP	4	0	3000	65		X	X		X			
Hooper et al. (2010)	50	Mood	IRAP	1	3	3000	80	48	X	X		X			
Roddy et al. (2010)	80	Weight	IRAP/IAT	1	3	3000	80	80	X	X		X	X	X	X
Stockwell et al. (2010)	17	Sex	IRAP	1	1	3000	80	?	X	X		X			
Timko et al. (2010)	168	Self-Esteem	IRAP	2	7	3000	65	84	X	X		X			
Vahey et al. (2010)	16	Smoking	IRAP	2	0	?	70	82	X						
Levin et al. (2011)	58	Procedural	MT-IRAP			2000	80	74		X		X	X		X
Hughes et al. (2011)	64	Attitudes	IRAP	1	2	3000	85	82	X			X		X	X
Roddy et al. (2011)	78	Body Image	IRAP/IAT/EMG	1	3	3000	80	82	X			X	X	X	X
Campbell et al. (2011)	60	Method	IRAP	1	1	3000	70	80	X	X	X	X			
Nicholson & Barnes-Holmes (2012)	40	Spider Fear	IRAP	1	2	2000	75	75	X	X	X	X			X
Parling et al. (2012)	34	Anorexia	IRAP	4	4	?	75	70-94	X	X		X			
Carpenter et al. (2012)	25	Cocaine	IRAP/Stroop	4		?	?	95	X	X		X			X
Hussey & Barnes-Holmes (2012)	30	Depression	IRAP	1	3	3000	80	83	X	X		X			
Timmins et al. (2011)	32	Sexual	IRAP	1	2	2000	80	100	X	X		X	X		X

Fourth and finally, although many of the assumptions made by the REC model have now been empirically confirmed, others await additional scrutiny. For instance, a systematic experimental analysis of the learning histories and current contextual variables critical to establishing, maintaining, and changing BIRRs and EERRs is clearly needed. Doing so will serve to clarify conceptual issues surrounding the specific properties of BIRRs and EERRs such as relational complexity, coherence and levels of derivation.

Conclusion

Relational Frame Theory makes two bold claims that differentiate it from many other approaches to psychological science. First, it argues that a defining feature of our species (complex language) is inherently relational in nature. Second, it proposes that we can understand, predict and influence relational responding using a purely functional perspective. As the current volume attests, the success of this functional approach is evident in the comprehensive, coherent, systematic and empirically driven literature that connects topics such as perspective taking, analogy, psychological disorders and intelligence under one unified theoretical umbrella. With respect to the study of implicit cognition, RFT has also served as the theoretical foundation out of which one functional methodology (IRAP) and a model (REC) have grown and flourished. Accumulating evidence provides support for the predictions of the REC model and, in particular, for the claim that "automatic" or "implicit" cognition is relational (propositional) rather than associative in nature. We fully recognize that much work remains if we are to better understand the specific properties of BIRRs and EERRs as well as their independent or interactive effects in predicting and influencing behavior. Nevertheless, we believe that the REC model in conjunction with the IRAP provides one viable functional means for doing so.

References

Amodio, D. M., & Ratner, K. G. (2011). A memory systems model of implicit social cognition. *Current Directions in Psychological Science, 20,* 143-148.

Banaji, M. R. & Heiphetz, L. (2010). Attitudes. In S. T. Fiske, D. T. Gilbert, & G.Lindzey (Eds.), *Handbook of Social Psychology (pp. 348-388).* New York: John Wiley & Sons.

Bargh, J. A. (1994). The Four Horsemen of automaticity: Awareness, efficiency, intention, and control in social cognition. In R. S. Wyer, Jr., & T. K. Srull (Eds.), Handbook of social cognition (2nd ed., pp. 1-40). Hillsdale, NJ: Erlbaum.

Barnes-Holmes, D., Barnes-Holmes, Y., Power, P., Hayden, E., Milne, R., & Stewart, I. (2006). Do you really know what you believe? Developing the Implicit Relational Assessment Procedure (IRAP) as a direct measure of implicit beliefs. *The Irish Psychologist, 32,* 169–177.

Barnes-Holmes, D., Barnes-Holmes, Y., Stewart, I., & Boles, S. (2010). A sketch of the implicit relational assessment procedure (IRAP) and the relational elaboration and coherence (REC) model. *The Psychological Record, 60,* 527-542.

Barnes-Holmes, D., Hayden, E., Barnes-Holmes, Y., & Stewart, I. (2008). The Implicit Relational Assessment Procedure (IRAP) as a response-time and event-related-potentials methodology for testing natural verbal relations: A preliminary study. *The Psychological Record, 58,* 497–515.

Barnes-Holmes, D., Murphy, A., Barnes-Holmes, Y., & Stewart, I. (2010). The Implicit Relational Assessment Procedure (IRAP): Exploring the impact of private versus public contexts and the response latency criterion on pro-White and anti-Black stereotyping among white Irish individuals. *The Psychological Record, 60,* 57–66.

Barnes-Holmes, D., Murtagh, L., Barnes-Holmes, Y., & Stewart, I. (2010). Using the Implicit Association Test and the Implicit Relational Assessment Procedure to measure attitudes towards meat and vegetables in vegetarians and meat-eaters. *The Psychological Record, 60,* 287-306.

Barnes-Holmes, D., Staunton, C., Barnes-Holmes, Y., Whelan, R., Stewart, I., Commins, S., Walsh, D., Smeets, P., & Dymond, S. (2004). Interfacing Relational Frame Theory with cognitive neuroscience: Semantic priming, The Implicit Association Test, and event related potentials. *International Journal of Psychology and Psychological Therapy, 4,* 215-240.

Barnes-Holmes, D., Staunton, C., Whelan, R., Barnes-Holmes, Y., Commins, S., Walsh, D., Stewart, I., Smeets, P. M., & Dymond, S. (2005). Derived stimulus relations, semantic priming, and event-related potentials: Testing a behavioral theory of semantic networks. *Journal of the Experimental Analysis of Behavior, 84,* 417–433.

Barnes-Holmes, D., Waldron, D., Barnes-Holmes, Y., & Stewart, I. (2009). Testing the validity of the Implicit Relational Assessment Procedure (IRAP) and the Implicit Association Test (IAT): Measuring attitudes towards Dublin and country life in Ireland. *The Psychological Record, 59,* 389–406.

Biglan, A., & Hayes, S. C. (1996). Should the behavioral sciences become more pragmatic? The case for functional contextualism in research on human behavior. *Applied and Preventive Psychology 5,* 47-57.

Campbell, C., Barnes-Holmes, Y., Barnes-Holmes, D., & Stewart, I. (2011). Exploring screen presentations in the Implicit Relational Assessment Procedure (IRAP). *International Journal of Psychology and Psychological Therapy, 11(3),* 377-388.

Carpenter, K. M., Martinez, D., Vadhan, N. P., Barnes-Holmes, D., & Nunes, E. V. (2012). Measures of attentional bias and relational responding are associated with behavioral treatment outcome for cocaine dependence. *The American Journal of Drug and Alcohol Abuse, 38,* 146.

Cassidy, S., Roche, B., & Hayes, S. C. (2011). A relational frame training intervention to raise Intelligence Quotients: A pilot study. *The Psychological Record, 61,* 173-198.

Chan, G., Barnes-Holmes, D., Barnes-Holmes, Y., & Stewart, I. (2009). Implicit attitudes to work and leisure among North American and Irish individuals: A preliminary study. *International Journal of Psychology and Psychological Therapy, 9,* 317-334.

Choma, B. L., & Hafer, C. L. (2009). Understanding the relation between explicitly and implicitly measured political orientation: The moderating role of political sophistication. *Personality and Individual Differences 47,* 964-967.

Cullen, C., Barnes-Holmes, D., Barnes-Holmes, Y., & Stewart, I. (2009). The Implicit Relational Assessment Procedure (IRAP) and the malleability of ageist attitudes. *The Psychological Record, 59,* 591–620.

Dawson, D. L., Barnes-Holmes, D., Gresswell, D. M., Hart, A. J. P., & Gore, N. J. (2009). Assessing the implicit beliefs of sexual offenders using the Implicit Relational Assessment Procedure: A first study. Sexual Abuse: *A Journal of Research and Treatment, 21,* 57–75.

De Houwer, J. (2006). What are implicit measures and why are we using them? In R.W. Wiers & A. W. Stacy (Eds.), *The handbook of implicit cognition and addiction* (pp. 11-28). Thousand Oaks, CA: Sage Publishers.

De Houwer, J. (2011). Why the cognitive approach in psychology would profit from a functional approach and vice versa. *Perspectives on Psychological Science, 6,* 202-209.

De Houwer, J., & Moors, A. (2012). How to define and examine implicit processes? In R. Proctor & E. J. Capaldi (Eds.), *Psychology of science: Implicit and explicit processes.* Oxford University Press.

Deutsch, R., & Gawronski, B. (2009). When the method makes a difference: Antagonistic effects on "automatic evaluations" as a function of task characteristics of the measure. *Journal of Experimental Social Psychology, 45,* 101-114.

Dixon, M. R., Rehfeldt, R. A., Zlomke, K. M., & Robinson, A. (2006). Exploring the development and dismantling of equivalence classes involving terrorist stimuli. *The Psychological Record 56,* 83-103.

Drake, C. E., Kellum, K. K., Wilson, K. G., Luoma, J. B., Weinstein, J. H., & Adams, C. H. (2010). Examining the Implicit Relational Assessment Procedure: Four pilot studies. *The Psychological Record, 60,* 81–100.

Fazio, R. H., Jackson, J. R., Dunton, B. C., & Williams, C. J. (1995). Variability in automatic activation as an unobtrusive measure of racial attitudes: A bona fide pipeline? *Journal of Personality and Social Psychology*, 69, 1013–1027.

Gavin, A., Roche, B., & Ruiz, M. R. (2008). Competing contingencies over derived relational responding: A behavioral model of the Implicit Association Test. *The Psychological Record*, 58, 427-441.

Gawronski, B., & Bodenhausen, G. V. (2011). The associative-propositional evaluation model: Theory, evidence, and open questions. *Advances in Experimental Social Psychology*, 44, 59-127.

Gawronski, B., & Creighton, L. A. (in press). Dual-process theories. In D. E. Carlston (Ed.), *The Oxford handbook of social cognition*. New York: Oxford University Press.

Gawronski, B., & Payne, B. K. (Eds.). (2010). *Handbook of implicit social cognition: Measurement, theory, and applications*. New York: Guilford Press.

Gifford, E. V., Hayes. S.C. (1999). Functional contextualism: A pragmatic philosophy for behavioral science. In: W. O'Donohue, & R. Kitchener, (Eds.), *Handbook of behaviorism*. New York: Academic Press.

Greenwald, A. G., Banaji, M. R., Rudman, L. A., Farnham, S. D., Nosek, B. A., & Mellott, D. S. (2002). A unified theory of implicit attitudes, stereotypes, self-esteem, and self-concept. *Psychological Review*, 109, 3–25.

Greenwald, A. G., McGhee, D. E., & Schwartz, J. K. L. (1998). Measuring individual differences in implicit cognition: The Implicit Association Test. *Journal of Personality and Social Psychology*, 74, 1464–1480.

Greenwald, A. G., Nosek, B. A., & Banaji, M. R. (2003). Understanding and Using the Implicit Association Test: I. An Improved Scoring Algorithm. *Journal of Personality and Social Psychology*, 85, 197-216.

Greenwald, A. G., Poehlman, T. A., Uhlmann, E., & Banaji, M. R. (2009). Understanding and using the Implicit Association Test: III. Meta-analysis of predictive validity. *Journal of Personality and Social Psychology*, 97, 17–41.

Hayes, S. C., Barnes-Holmes, D., & Roche, B. (Eds.). (2001). *Relational Frame Theory: A Post-Skinnerian account of human language and cognition*. New York: Plenum Press.

Hooper, N., Villatte, M., Neofotistou, E., & McHugh, L. (2010). An implicit versus an explicit measure of experimental avoidance. *International Journal of Behavior Consultation and Therapy*, 6(3), 233-245.

Hughes, S., & Barnes-Holmes, D. (2011). On the formation and persistence of implicit attitudes: New evidence from the Implicit Relational Assessment Procedure (IRAP). *The Psychological Record*, 61, 391–410.

Hughes, S., Barnes-Holmes, D., & De Houwer, J. (2011). The dominance of associative theorizing in implicit attitude research: Propositional and behavioral alternatives. *The Psychological Record*, 61, 465–496.

Hull, C. L. (1943). *Principles of behavior*. New York: Appleton-Century-Crofts.

Hussey, I., & Barnes-Holmes, D. (2012). The IRAP as a measure of implicit depression and the role of psychological flexibility. *Cognitive and Behavioral Practice, 19*, 573.

Leslie, J., Tierney, K. J., Robinson, C. P., Keenan, M., Watt, A. & Barnes, D. (1993). Differences between clinically anxious and non-anxious subjects in a stimulus equivalence training task involving threat words. *The Psychological Record, 43*, 153-161.

Levin, M. E., Hayes, S. C., & Waltz, T. (2010). Creating an implicit measure of cognition more suited to applied research: A test of the Mixed Trial—Implicit Relational Assessment Procedure (MT-IRAP). *International Journal of Behavioral Consultation and Therapy, 6*, 245-261.

Maison, D., Greenwald, A. G., & Bruin, R. H. (2004). Predictive validity of the Implicit Association Test in studies of brands, consumer attitudes, and behavior. *Journal of Consumer Psychology, 14*, 405–415.

McGlinchey, A., Keenan, M., & Dillenburger, K. (2000). Outline for the development of a screening procedure for children who have been sexually abused. *Research on Social Work Practice, 10*, 722-747.

McKenna, I. M., Barnes-Holmes, D., Barnes-Holmes, Y., & Stewart, I. (2007). Testing the fake-ability of the Implicit Relational Assessment Procedure (IRAP): The first study. *International Journal of Psychology and Psychological Therapy, 7*, 253-268.

Merwin, R. M. & Wilson, K. G. (2005). Preliminary Findings on the Effects of Self-Referring and Evaluative Stimuli on Stimulus Equivalence Class Formation. *The Psychological Record, 55*, 561-575.

Nicholson, E., & Barnes-Holmes, D. (2012b). The Implicit Relational Assessment Procedure (IRAP) as a measure of spider fear. *The Psychological Record, 62*, 263.

Nicholson, E., & Barnes-Holmes, D. (2012a). Developing an implicit measure of disgust propensity and disgust sensitivity: Examining the role of implicit disgust propensity and sensitivity in obsessive-compulsive tendencies. *Journal of Behavior Therapy and Experimental Psychiatry, 43*, 922-930.

Nosek, B. A. (2007). Implicit-explicit relations. *Current Directions in Psychological Science, 16*, 65-69.

Nosek, B. A., & Banaji, M. R. (2001). The go/no-go association task. *Social Cognition, 19*, 625–664.

Nosek, B. A., Hawkins, C. B., & Frazier, R. S. (2011). Implicit social cognition: From measures to mechanisms. *Trends in Cognitive Sciences, 15*, 152-159.

Olson, M. A., & Fazio, R. H. (2009). Implicit and explicit measures of attitudes: The perspective of the MODE model. In R. E. Petty, R. H. Fazio, & P. Briñol (Eds.), *Attitudes: Insights from the new implicit measures* (pp. 19–63). New York: Psychology Press.

O'Reilly, A., Roche, B., Ruiz, M., Tyndall, I., & Gavin, A. (2012). The Function Acquisition Speed Test (FAST): A behavior-analytic implicit test for assessing stimulus relations. *The Psychological Record, 62*, 507.

O'Toole, C., & Barnes-Holmes, D. (2009). Three chronometric indices of relational responding as predictors of performance on a brief intelligence test: The importance of relational flexibility. *The Psychological Record, 59*, 119–132.

O'Toole, C., Barnes-Holmes, D., & Smyth, S. (2007). A derived transfer of functions and the Implicit Association Test. *Journal of the Experimental Analysis of Behavior, 88*, 263-283.

Parling, T., Cernvall, M., Stewart, I., Barnes-Holmes, D., & Ghaderi, A. (2012). Using the Implicit Relational Assessment Procedure to compare implicit pro-thin/anti-fat attitudes of patients with anorexia nervosa and non-clinical controls. *Eating Disorders: The Journal of Treatment and Prevention, 20*, 127.

Payne, B. K., Burkley, M., & Stokes, M. (2008). Why do implicit and explicit attitude tests diverge? The role of structural fit. *Journal of Personality and Social Psychology, 94*, 16-31.

Payne, B. K., Cheng, S. M., Govorun, O., & Stewart, B. D. (2005). An inkblot for attitudes: Affect misattribution as implicit measurement. *Journal of Personality and Social Psychology, 89*, 277–293.

Payne, B. K., & Gawronski, B. (2010). A history of implicit social cognition: Where is it coming from? Where is it now? Where is it going? In B. Gawronski & B. K. Payne (Eds.), *Handbook of implicit social cognition: Measurement, theory, and applications* (pp. 1–15). New York: Guilford Press.

Perugini, M., Richetin, J., & Zogmaister, C. (2010). Prediction of behavior. In B. Gawronski, & K. Payne (Eds.), *Handbook of Implicit Social Cognition: Measurement, Theory, and Applications* (255-277). New York: Guilford.

Petty, R. E., Briñol, P., & De Marree, K. G. (2007). The meta-cognitive model (MCM) of attitudes: Implications for attitude measurement, change, and strength. *Social Cognition, 25*, 657–686.

Power, P. M., Barnes-Holmes, D., Barnes-Holmes, Y., & Stewart, I. (2009). The Implicit Relational Assessment Procedure (IRAP) as a measure of implicit relative preferences: A first study. *The Psychological Record, 59*, 621–640.

Roche, B., Ruiz, M., O'Riordan, M., & Hand, K. (2005). A relational frame approach to the psychological assessment of sex offenders. In M. Taylor & E. Quayle (eds.), *Viewing child pornography on the internet: Understanding the offense, managing the offender, and helping the victims* (pp. 109–125). Dorset, UK: Russell House Publishing.

Roddy, S., Stewart, I., & Barnes-Holmes, D. (2010). Anti-fat, pro-slim, or both? Using two reaction time based measures to assess implicit attitudes to the slim and overweight. *Journal of Health Psychology, 15*, 416-425.

Roddy, S., Stewart, I., & Barnes-Holmes, D. (2011). Facial reactions reveal that slim is good but fat is not bad: Implicit and explicit measures of body-size bias. *European Journal of Social Psychology, 41(6)*, 688-694.

Rydell, R. J., & McConnell, A. R. (2006). Understanding implicit and explicit attitude change: A systems of reasoning analysis. *Journal of Personality and Social Psychology, 91*, 995–1008.

Schwartz, M. B., Vartanian, L. R., Nosek, B. A., & Brownell, K. D. (2006). The influence of one's own body weight on implicit and explicit anti-fat bias. *Obesity, 14,* 440-447.

Snowden R. J., Wichter, J., & Gray N. S. (2008). Implicit and explicit measurements of sexual preference in gay and straight men: A comparison of priming techniques and the implicit association task. *Archives of Sexual Behavior, 37,* 558-565.

Stockwell, F. M., Walker, D. J., & Eshleman, J. W. (2010). Measures of implicit and explicit attitudes toward mainstream and BDSM sexual terms using the IRAP and questionnaire with BDSM/Fetish and student participants. *The Psychological Record, 60,* 307–324.

Strack, F., & Deutsch, R. (2004). Reflective and impulsive determinants of social behavior. *Personality and Social Psychology Review, 8,* 220–247.

Teachman, B. A. (2007). Evaluating implicit spider fear associations using the go/no-go association task. *Journal of Behavior Therapy and Experimental Psychiatry, 38,* 156–167.

Teachman, B. A., Gregg, A. P., Woody, S. R. (2001). Implicit associations for fear-relevant stimuli among individuals with snake and spider fears. *Journal of Abnormal Psychology, 110,* 226–35.

Timko, C. A., England, E. L., Herbert, J. D., & Forman, E. M. (2010). The Implicit Relational Assessment Procedure as a measure of self-esteem. *The Psychological Record, 60,* 679–698.

Timmins, L., Barnes-Holmes, D., & Cullen, C. (2011). Measuring implicit sexual response biases to nude male and female pictures in gay and heterosexual men. Unpublished manuscript.

Vahey, N. A., Barnes-Holmes, D., Barnes-Holmes, Y., & Stewart, I. (2009). A first test of the Implicit Relational Assessment Procedure (IRAP) as a measure of self-esteem: Irish prisoner groups and university students. *The Psychological Record, 59,* 371–388.

Vahey, N., Boles, S., & Barnes-Holmes, D. (2010). Measuring adolescents' smoking related social identity preferences with the Implicit Relational Assessment Procedure (IRAP) for the first time: A starting point that explains later IRAP evolutions. *International Journal of Psychology and Psychological Therapy, 10,* 453-474.

Vartanian, L. R., Herman, P. C., & Polivy, J. (2005). Implicit and explicit attitudes toward fatness and thinness: The role of the internalization of societal standards. *Body Image, 2,* 373-381.

Watson, J. B. (1924/1925). *Behaviorism.* New York: The People's Institute Publishing Company.

Watt, A. W., Keenan, M., Barnes, D., & Cairns, E. (1991). Social categorization and stimulus equivalence. *The Psychological Record, 41,* 33–50.

Whelan, R. (2008). Effective analysis of reaction time data. *The Psychological Record, 58,* 475-82.

Wittenbrink, B., Judd, C. M., & Park, B. (1997). Evidence for racial prejudice at the implicit level and its relationship to questionnaire measures. *Journal of Personality and Social Psychology, 72*, 262–274.

CHAPTER SIX

Advances in Research on Deictic Relations and Perspective-Taking

Yvonne Barnes-Holmes

Mairéad Foody

Dermot Barnes-Holmes

National University of Ireland, Maynooth,

Louise McHugh

University College Dublin

If there is any secret of success, it lies in the ability to get the other person's point of view and see things from his angle as well as your own. (Henry Ford)

According to Relational Frame Theory (RFT), repertoires of derived relational responding are essential for all stages of development because they are a defining feature of human verbal behavior. Indeed, relational framing appears to underpin many of the basic phenomena that comprise human language and cognition, including: naming; storytelling; metaphor; grammar; humor; empathy; and perspective-taking. As a result, the implications of relational framing for human development and for the education of developmentally delayed populations are immense.

Perspective-taking, in terms of deictic relations, has become one of the most widely studied areas of relational responding, as dictated by RFT (e.g., McHugh, Barnes-Holmes, Barnes-Holmes, 2004a). In the first section of the current chapter, we summarize the RFT definition of perspective-taking, as provided in the 2001 account, as well as briefly reviewing the empirical evidence available at that time. Thereafter, the primary focus of the chapter concerns the articulation of what we have learned about deictic relations since then, and a summary of the empirical evidence on relational perspective-taking that now exists. This evidence has continued to accumulate, as expected, from studies on both development and education with typically-developing children, as well as those with developmental disabilities (e.g., autism) and other psychiatric diagnoses (e.g., Asperger's syndrome). However, recent developments have also examined the potential role of relational perspective-taking in other diagnostic samples, such as individuals with schizophrenia, and have explored the role of perspective-taking in one's sense of self. The current chapter attempts to give readers a flavor of the breadth of application that has been achieved by RFT's account of perspective-taking.

Defining Perspective-Taking

In spite of its long-established centrality in typical and atypical patterns of development (Leslie, 1987) and numerous theoretical frameworks (e.g., Kohlberg, 1976), one mainstream cognitive account has dominated academic thinking about perspective-taking for almost 20 years. And, behavioral psychologists traditionally devoted relatively little attention to these phenomena.

Theory of Mind (ToM)

The basic proposition in the ToM approach to perspective-taking is that it involves, as the term implies, the development of a theory or understanding about one's own mind and the mind of others (see Baron-Cohen, Tager-Flusberg, & Cohen, 2000). The ToM approach comprises a somewhat universal, invariant, and increasingly complex five-stage account that ranges from simple visual perspective-taking to predicting actions on the basis of false belief (Howlin, Baron-Cohen, & Hadwin, 1999). According to this view, perspective-taking abilities typically begin to emerge between the ages of two and five (Baron-Cohen et al., 2000), and are characteristically deficient or delayed in individuals with autism (Frith, Happé, & Siddons, 1994). There have also been suggestions that these skills show atypical development in individuals with psychosis (Brüne, 2005). Indeed, it is well-established that perspective-taking deficits are associated with significant impairments in a range of areas including general communication skills (e.g., Tager-Flusberg & Anderson, 1991).

RFT's (2001) Conceptualization of Perspective-Taking as Relational Responding

For RFT, the skills involved in perspective-taking are an inherently relational form of verbal behavior, and the specific relations involved were initially referred to as "deictic relations". Although these are not, at the level of process, different from any other type of relational frames, they do appear to be particularly complex, because they have no obvious non-arbitrary counterparts. This complexity suggests that, in the natural sequence of development, the majority of relational frames (e.g., comparison, distinction, and opposition) will be established prior to the emergence of perspective-taking.

The deictic relational frames comprise the spatial relations of I-YOU and HERE-THERE, as well as the temporal relations of NOW-THEN. And a great deal of behavior-analytic literature supports the importance of understanding spatial and temporal dimensions in the normal course of human development (e.g., O'Hora, Peláez, & Barnes-Holmes, 2005). Indeed, it is easy to see the role of these relations in many common verbal interactions with others (e.g., "What are *you* doing *now?*", "*Where* shall *we* meet *later?*", and "If only *I* had known *then* what *I* know *now*"). Although the form of questions such as these is often identical, the physical and temporal environment is constantly changing. For example, if you say "now" meaning three o'clock and you say "now" later in reference to nine o'clock, the time has changed, although the perspective (i.e., the

I) from which time is observed has remained constant. So, for RFT, the relational properties of I vs. YOU, HERE vs. THERE, and NOW vs. THEN become constants, against which environments that are constantly changing in terms of time and space can be understood, categorized, and communicated about. As is the case for all relational frames, the properties of the perspective-taking relations may be present in the use of specific words, such as "I", "you", and "here", but our conversations include alternative words which also specify these relations (e.g., "It is one o'clock and I am at work (HERE and NOW), but Mary (YOU) is still in the cafe" (THERE and NOW).

During language and cognitive development, the relational properties of perspective-taking are believed to be abstracted through learning to talk about your perspective in relation to others (McHugh, Barnes-Holmes, O'Hora, & Barnes-Holmes 2004), and across multiple exemplars this abstraction generates the constancy that characterizes your perspective. However, because, for RFT, this is generalized relational behavior, competence and flexibility in other relational frames would likely serve as prerequisites for relational perspective-taking. For example, you could not know that your perspective was different from your friend's if you could not respond in accordance with distinction relations.

Empirical Evidence on Relational Perspective-Taking Available in 2001

In comparison to the burgeoning body of empirical evidence on deictic relations that exists today, there was little research available in the original RFT book. Indeed, the evidence reviewed therein focused on related research on: (1) the transformation of functions in accordance with symmetry; (2) establishing the frames of more-than, less-than, and opposite; and (3) the utility of multiple-exemplar training in establishing these frames. Although these studies did not provide *direct* evidence that perspective-taking comprises relational responding, they demonstrated that many core features of human development are based on derived relations. Furthermore, those studies provided support for the ease with which young children acquire relational repertoires and may be trained to do so. As an operant theory of language and cognition, it was critical for RFT that these two effects were demonstrated, and a number of studies since have lent additional support to the role of derived relations in this regard (see below).

In the 10 years since the first RFT book, the literature that has enhanced our understanding of perspective-taking as relational responding has comprised three key strands. First, there are numerous developmental studies, with both typically-developing and other children, which have identified repertoires of relational responding that appear to be prerequisites for perspective-taking. Second, there are studies which examine perspective-taking more directly and have assessed the capabilities of various groups of individuals on existing protocols of deictic responding. Several studies in this strand have also investigated the use of these protocols as training tools with typical and atypical populations. Third, there is a small but growing body of research on the relationship between deictic relations and the self, with some studies focusing on the therapeutic implications of this relationship, and others on the role of perspective-taking in forms of psychopathology, such as schizophrenia.

Evidence on Precursors to Relational Perspective-Taking

Since 2001, a growing body of evidence has provided further support to the importance of repertoires of derived relational responding in the natural course of development. For RFT, relational repertoires are cumulative in the sense that development in one family of relational frames impacts upon the emergence of another. For example, you cannot easily derive opposition relations if you have no previous experience with distinction relations (because opposition is a type of distinction). Hayes and Wilson (1996) summarized this overlap as follows:

> It does seem likely that once the most basic relational unit is established through training in mutual and combinatorial entailment, relatively fewer trained instances of combinatorial entailment will be needed to build out this relational response. Were it not true, every level of relational complexity (e.g., with larger and larger sets of related stimuli) might have to be arduously trained. (p.227)

Although it would be an empirical issue to precisely determine that specific relational repertoires necessarily precede others, numerous studies have examined early relational repertoires that most likely constitute the basis of relational perspective-taking. The available evidence on these is summarized briefly below.

The relational frame of coordination. The derivation of coordination relations appears to be one of the earliest forms of relational responding and dominates language development in infants (Hayes et al., 2001). Word-object and object-word relations are perhaps the most common examples of mutual entailment, which develops into language in part through combinatorial entailment. Luciano, Gómez-Becerra, and Rodríguez-Valverde (2007) demonstrated combinatorial entailment in a 19-month old infant, the earliest age at which this type of derivation has been empirically demonstrated. Another study by O'Connor, Rafferty, Barnes-Holmes, and Barnes-Holmes (2009) reported a relationship between low verbal competence and limited combinatorial entailment, and facilitated the latter in 15 children diagnosed with autism (ages 5-6).

The relational frame of opposition. The derivation of opposition relations involves arbitrarily applying the relational cue "is the opposite of" (or something similar) to any stimuli or event. From a developmental perspective, it is most likely that frames of coordination are necessary for the emergence of opposition relations because the combinatorially entailed relations within opposition are coordinated. For example, if A is the opposite of B and B is the opposite of C, then A and C are coordinated. Barnes-Holmes, Barnes-Holmes, and Smeets (2004) successfully generated repertoires of relational responding in accordance with opposite in typically-developing children, aged between 4 and 6 years.

The relational frame of distinction. Behaving in accordance with the frame of distinction involves responding to differences among stimuli and arbitrarily applying the relational cue "is different from", or something similar. While the contextual cues for opposition allow the precise relations to be derived, frames of distinction often do not specify the relevant dimensions. That is, you can only be opposite to something in one way, but you can be different from something in many ways.

The relational frame of comparison. Similar to distinction relations, comparison relations require the specification of the dimension along which two stimuli are to be compared. For example, of a pair of coins, one may be worth more, and so the comparative dimension is value. Barnes-Holmes, Barnes-Holmes, Smeets, Strand, and Friman (2004) successfully generated repertoires of relational responding in accordance with more-than and less-than in typically-developing children, aged between 4 and 6 years. Berens and Hayes (2007) replicated these findings with four other children (aged 4-5 years) and demonstrated generalization across stimuli and trial-type. Furthermore, Gorham, Barnes-Holmes, Barnes-Holmes, and Berens (2009) generated repertoires of relational responding in accordance with more-than and less-than in five typically-developing children and three children with autism aged between 6 and 9 years old. As expected, the children with autism required more extensive training than their typically-developing peers.

The relational frame of hierarchy. The relational frame of hierarchy involves responding in accordance with contextual cues, such as "contains", "is a member of", or "belongs to". Kinship relations are perhaps the most common example. For instance, if Tom is the father of Ann and John, you can derive that Ann and John are siblings. In one study by Griffee and Dougher (2002), hierarchical categorization was established in an adult population using conditional discrimination training, stimulus generalization, and stimulus equivalence across two experiments. However, there are no published studies to date which have investigated hierarchical relations from an RFT perspective.

In the section above, we have outlined the core relational frames which appear to be well established prior to the emergence of the perspective-taking relations. However, two obvious questions emerge. First, what is the sequence of the natural emergence of these families of relations (e.g., to what extent do they precede, or overlap with, each other)? Second, how can we know that they are necessary for the development of perspective-taking?

In the natural course of development, it would appear that coordination relations emerge first, because the bulk of early language interactions involve the establishment of coordination relations between word, objects, pictures, etc. Even prior to this development of arbitrary relations (i.e., a word is not physically the same as the object to which it

133

relates, hence the relation is arbitrary), infants demonstrate non-arbitrary identity matching or stimulus reflexivity. For RFT, this non-arbitrary history is essential for the subsequent emergence of arbitrary coordination relations, such as relating words and objects. In the context of arbitrary relations, it also makes sense that a child learns what is the same before she learns what is not the same, hence coordination relations, at least to some extent, likely precede distinction or opposition relations. However, that is not to say that all coordination relations *must* be fully established and operating at high levels of relational flexibility and complexity before acquisition of distinction relations, for example, can begin. It is more likely that once coordination relations are well-established, exemplar training with other types of relations naturally begins. Similar issues emerge with regard to whether distinction relations precede opposition relations, and in normal development they most likely emerge in tandem.

For RFT, it does seem likely that coordination, distinction and opposition relations, on the whole, precede hierarchical and deictic relations, and perhaps hierarchical relations even precede deictic ones. Hierarchical relations involve the concept of containment, along various dimensions, hence at one level hierarchically related events are both coordination (i.e., of the same family), but are distinct (i.e., they are not at the same level as each other). The suggestion that hierarchical relations probably precede deictic relations rests on the fact that deictic relations have no clear non-arbitrary basis on which they can be established (i.e., your perspective is always changing) and this makes their emergence more complex.

Evidence Using the Relational Perspective-Taking Protocol

The first RFT study on perspective-taking was reported by Barnes-Holmes (2001) and involved the development of a testing and training protocol for assessing and facilitating/establishing the three deictic relations (I-YOU, HERE-THERE, and NOW-THEN) in young children. The protocol comprised 256 trials or relational tasks that spanned three levels of relational complexity, referred to as: simple relations, reversed relations, and double reversed relations.

Consider the following simple I-YOU trial. Participants are asked "If [hypothetically speaking] I (experimenter) have a red brick and YOU (participant) have a blue brick: Which brick do I have? Which brick do YOU have?" This requires participants to respond under the contextual control of an if-then relation and in accordance with I-YOU relations, and the answer simply is as stated (experimenter would have red and participant would have blue). The trial is denoted as I-YOU because it explicitly targets I-YOU relations and is "simple" in terms of the level of relational complexity because these relations are not reversed (i.e., they are as stated in the question). Naturally, I-YOU relations are the first of the simple relations to be established, because they form the basis of the spatial and temporal relations, which require that the listener can arbitrarily distinguish between the two perspectives (i.e., between "I" the speaker and "You" the listener).

Simple HERE-THERE trials are similar in terms of operating at the same simple level of relational complexity, but these also explicitly target HERE-THERE or spatial relations. (e.g., "I am sitting here on the blue chair, and you are sitting there on the black chair. Where am I sitting? Where are you sitting?"). It is important to note, however, that simple HERE-THERE relations include I-YOU relations because it would be impossible to specify a perspective from a particular location if there was no perspective from which to operate. Furthermore, it is important to emphasize that I is always located HERE and YOU is always located THERE when each of these relations are operating at a simple level of relational complexity. In this way, I is always coordinated with HERE (and distinct or opposite to THERE), just as YOU is always coordinated with THERE (and distinct or opposite to HERE).

The third type of simple relations are temporal and specify NOW-THEN relations (e.g., "Yesterday I was watching television, today I am reading. What was I doing then? What am I doing now?"). Of course, these are perhaps more abstract at one level than spatial relations because, for example, "yesterday" has no non-arbitrary properties. Nonetheless, these relations are still simple in terms of relational complexity. A further distinction between these and the spatial relations above concerns the presentation of only one perspective. Look at the example again and note that on simple NOW-THEN trials, participants

135

are only asked about what "I" was doing yesterday and am doing now, and "YOU" is not mentioned. This limitation ensures that all of the relations can be specified by the speaker. For instance, if you are told "Yesterday, I was reading; today you are watching television," you could not derive, without additional information, what I am doing today or what you were doing yesterday. For that reason, all NOW-THEN trials present either I-only or YOU-only relations from the I-YOU pair. This limitation affects temporal relations only.

All three deictic relations can be reversed at the level of relational complexity and this demonstrates the arbitrariness of the relations. Consider the reversed I-YOU trial in which participants are asked "If I had a red brick and you had a blue brick. If I were YOU and YOU were me. Which brick would I have? Which brick would you have?" As the name implies, the reversal of the I-YOU relation arbitrarily switches the perspective, such that I have your perspective and you have mine. Hence, the correct response is the reversal of the instructions (i.e., I would have the blue brick and you would have the red brick).

The spatial HERE-THERE relations may be reversed in the same way (e.g., "If I am sitting here on the blue chair and you are sitting there on the black chair. If HERE was THERE and THERE was HERE"). Technically, this is an example in which the spatial relation has been reversed, but the more implicit I-YOU relation remains simple (i.e., because it was not explicitly reversed). However, in any instance of correct responding, one cannot know whether the answer is based on the reversal of the I-YOU relations or a reversal of the HERE-THERE relations. In other words, if the HERE-THERE relations are coordinated, respectively with I-YOU relations, one could be reversing either. However, the most important point to note is that responding correctly can be done on the basis of reversing *either, not both* of the perspective or spatial relations (see below for double reversals). Put simply, there are only two perspectives (mine and yours) and hence only one way of switching them around.

The temporal NOW-THEN relations may also be reversed (e.g., "Yesterday I was reading, today I am watching TV. If NOW was THEN and THEN was NOW"). However, unlike the reversed spatial relations, and because of the single perspective (I or YOU), responding correctly can only be on the basis of the reversed temporal relation. That is, there is no switch in perspectives across people, just a switch in the temporal dimension of a single perspective. This switch also demonstrates the arbitrariness of one's perspective, because although I is always coordinated with NOW, it can arbitrarily be coordinated with THEN (because I was also I THEN). But obviously when I was operating my perspective from then, it was NOW at that time.

Two simultaneously reversed deictic relations define the highest level of relational complexity and are denoted as *double reversals*. There are only two types of double reversals: I-YOU/HERE-THERE and HERE-THERE/NOW-THEN. There is no such trial as an I-YOU/NOW-THEN double reversal because the temporal relations specify only one perspective. Consider an I-YOU/HERE-THERE double reversed trial that states "I am sitting here on the blue chair and you are sitting there on the black chair. If I were YOU and YOU were me, and if HERE was THERE and THERE was HERE: Where would you be sitting? Where would I be sitting?"). This is basically a simple two-step process in which the perspectives are first switched by the I-YOU reversal and are then switched *back* in the HERE-THERE reversal (vice versa is also possible). In any case, a correct answer is as stated in the trial because one goes back to one's original perspective.

Consider a HERE-THERE/NOW-THEN double reversal trial in which a participant is asked "Yesterday I was sitting there on the blue chair, today I am sitting here on the black chair. If HERE was THERE and THERE was HERE, and if NOW was THEN and THEN was NOW: Where would I be sitting now? Where would I be sitting then?"). Once again, the perspectives are switched by the I-YOU reversal and then switched *back* in the NOW-THEN reversal (or vice versa), so that a correct answer is as stated in the trial (i.e., yesterday on blue, today on black).

Empirical Evidence Using the Relational Perspective-Taking Protocol with Typically-Developing Samples

The first full-relational perspective-taking protocol employed by Barnes-Holmes (2001) examined only I-YOU and HERE-THERE relations with a typically-developing seven-year old girl and four-year old boy. The baseline protocol data indicated that the young girl could derive all of the target relations, except the two double reversals, while the young boy failed on all types of relations at all levels of relational complexity. Across multiple exemplars, the older child required explicit training to complete the I-YOU/HERE-THERE double reversal and demonstrate the target performances. The younger child required a total of 55 sessions of testing and training and demonstrated the target performances up to the reversals. A subsequent perspective-taking study by McHugh et al. (2004a) explored the potential relationship between age/verbal sophistication and competence on an abbreviated 62-trial version of the original protocol. Participants were selected from five age groups; 3-5 years (early childhood), 6-8 years (middle childhood), 9-11 years (late childhood), 12-14 years (adolescence) and 18-30 years (adulthood). Their competencies on the three deictic relations and on the three levels of relational complexity were then systematically compared.

In general, adults produced the lowest number of errors across trial-types while the early childhood group produced the highest. This outcome suggested that errors increased as a function of relational complexity. All five age groups produced significantly more errors on all three types of reversed relations than on their matching simple relations (e.g., more errors on reversed I-YOU than simple I-YOU relations). Significant differences were also recorded for all participants in which there were more errors on: simple NOW-THEN than HERE-THERE relations; reversed HERE-THERE than reversed I-YOU relations; and reversed NOW-THEN than reversed I-YOU relations, suggesting that the distinctions made between the different types of relations had some validity. Two subsequent control studies demonstrated that the children's poor performances were not influenced by length of instructions and that the testing format did not influence the previous adult outcomes (see also McHugh, Barnes-Holmes, O'Hora, & Barnes-Holmes, 2004).

In an attempt to systematically compare the relational perspective-taking protocol with more traditional ToM tasks, Weil, Hayes, and Capurro (2011) presented both types of protocols to three typically-developing children (aged 4-5 years). In short, explicit training on the relational perspective-taking protocol, especially on reversed and double reversed relations, facilitated improvements on ToM test outcomes. Interestingly, one of the two children required more training in the reversals (8 trials) than on the double reversals (3 trials).

Heagle and Rehfeldt (2006) presented an automated 57-trial version of the relational perspective-taking protocol to three typically-developing children aged 6-11 years, and tested for generalization to real-world conversations. For example, consider the following generalized simple I-YOU trial: "I have a hamburger and you have a grilled cheese." One child required explicit training at all three levels of relational complexity, while the remaining two required training on the reversals (maximum 5 trials) and double reversals only (maximum 3 trials). Interestingly, all three children required more training on the reversals than double reversals. Most importantly, the researchers demonstrated the generalization of these skills to real-world conversations.

The previous study was replicated by Davlin, Rehfeldt, and Lovett (2011) with three additional typically-developing children, aged 6-11 years. However, these researchers modified the perspective-taking protocol by replacing YOU with the perspective of a story character (e.g., "You are reading books with me. Cinderella is doing chores. What are you doing? What is Cinderella doing?"). Once again, all three children demonstrated the target performances after substantive training on the protocol, although most training here was required on the double reversals (8 trials compared to 4 trials for reversals).

Empirical Evidence Using the Relational Perspective-Taking Protocol with Non-Typically-Developing Samples

In spite of nearly a decade of RFT research into deictic relations in typically-developing children, attempts to investigate or establish this behavior in populations with suspected deficits in these abilities are only starting to emerge. This research is discussed briefly below.

Gore, Barnes-Holmes, and Murphy (2010) examined putative correlations between performances on a 34-trial version of the relational perspective-taking protocol and varying degrees of intellectual disability with a sample of 24 adults. The results showed significant correlations between perspective-taking competence and: verbal mental age (as measured on the British Picture Vocabulary Scale, BPVS-II); verbal performance (as measured on a subscale of the Wechsler Abbreviated Scale of Intelligence, WASI); and full-scale IQ (as measured on the total WASI score).

Research by Rehfeldt, Dillen, Ziomek, and Kowalchuk (2007) investigated whether nine males with autism (aged 6-13 years) displayed more perspective-taking deficits than nine typically-developing peers. The results demonstrated that both groups produced nearly significantly weaker performances on reversed relations, relative to simple relations. However, the children with autism also produced more errors on the reversed relations than the typically-developing children. The same researchers also investigated whether competence in relational perspective-taking correlated with scores on two standardized tests of autism, namely the Vineland Adaptive Behavior Scales (VABS)-Interview Edition and the Social Communication Questionnaire (SCQ)-Current Form. The data indicated a correlation between performances on the NOW-THEN reversals and the Daily Living subscale of the VABS.

Taken together, the studies above demonstrate the research interest generated by RFT's perspective-taking protocol as a developmental and educational tool. The findings also show the relative ease with which young typically-developing children acquire these abilities.

Deictic Relations, True/False Belief, and Deception

According to the ToM account, responding on the basis of false belief and deception involves extensions of perspective-taking, and hence the latter must precede the former (Baron-Cohen et al., 2000). And findings from several RFT studies do indicate strong functional overlap among the deictic relations, false belief, and deception.

McHugh, Barnes-Holmes, Barnes-Holmes, and Stewart (2006) investigated the role of deictic relations in true and false belief among 40 participants grouped by age in a similar manner to McHugh et al. (2004a). The basic protocol comprised six tasks which emphasized the spatial and temporal relations, with false belief introduced in terms of "logical not". The six trial-types were referred to as: HERE; THERE; NOT HERE; NOT THERE; BEFORE NOW; and AFTER NOW. Consider a HERE trial in which participants were asked "If YOU put the pencil in the sweet box and I am HERE. What would I think is in the sweet box? What would YOU think is in the sweet box?" Participants were given the option of selecting either "Pencil" or "Sweets" as answers to either question.

In the language of RFT, responding on the basis of true belief involves the deictic relations, whereas responding on the basis of false belief is more complex because it involves combining the deictic relations with logical not. Consider the NOT HERE trial "If you put the pencil in the sweet box and I am not here. What would you think is in the sweet box? What would I think is in the sweet box?" This trial denotes that "I" am responding on the basis of false belief because "I" did not see you place the sweet in the sweet box, and thus "I" cannot know that you did so. Alternatively, if I had seen you do this, I could respond to that as a true belief, because I can see it here and now. In line with this account, the results of McHugh et al. (2006) demonstrated a developmental trend in terms of the total number of correct responses produced by participants across the five age groups, and more importantly demonstrated the relationship between age and competency on the false belief trials, in particular. In short, although all groups showed relative competence on the true belief trials, the older groups produced better performances on false belief trials than their younger counterparts.

Other research by McHugh, Barnes-Holmes, Barnes-Holmes, Stewart, & Dymond (2007) explored the role of deictic relations in deception, as an even more complex type of false belief. Once again, the researchers investigated a possible developmental profile for deictic relations in deception skills with 40 participants across five age groups. The deception protocol contained four basic trial-types, which involved deictic relations, logical not, and a new type of trial that involved one perspective knowing what another perspective knows (e.g., "I know what you know").

Consider the first-order positive trial, "If I have a teddy and I want you to find it, where should I put the teddy?" Given a choice of "toy box" or "refrigerator", the correct answer is "toy box" because I can take your perspective and you will expect a teddy to be in a toy box. At the level of relational complexity, therefore, this is an I-YOU reversal, because I can take your perspective, as if you and I had switched perspectives (e.g., "if I was you"). However, the term first-order is used here to distinguish these from the more complex trials, denoted as second-order. First-order trials, such as the example above, were also referred to as positive to denote the absence of logical not, which, in this protocol, indicated deception (just as logical not had been used previously to indicate false belief).

Now consider the second-order positive trial "If you have a teddy and you know that I know you're trying to hide it from me, where should you hide the teddy?" Given a choice of "toy box" or "refrigerator", the correct again answer is "toy box" because I can take your perspective and I know that you know this, so that you will expect me to hide it in the refrigerator, but because I am trying to hide it from you I will put it in the toy box. Relationally, this is like a double reversal, in which you know what I know (I-YOU reversal) and I know that you know this (a second reversal). Hence, the correct answer returns to the original location because the reversals cancel each other out.

In this protocol, deception was denoted by the inclusion of logical not, as applied to both first- and second-order trial-types. Consider the first-order negative trial, "If you have a teddy and you don't want me to find it, where should you put the teddy?" Given a choice of "toy box" or "refrigerator", the correct answer is "refrigerator" because you can take my perspective and I will expect a teddy to be in a toy box, but because of the deception, you are not going to put it there. Again, this is an I-YOU reversal, but the logical not, which denotes the deception, changes the answer from toy box to refrigerator. The application of logical not applies in the same way to the second-order trials, and thus simply switches the answer around.

The results from McHugh et al. (2007) demonstrated increased competencies in line with age. Furthermore, greater accuracies were recorded on first-order than second-order trials for all groups, and on positive vs. negative trials, but only among the younger groups of participants. Taken together, the growing body of research on the deictic relations provides considerable support for their role in what is traditionally referred to as perspective-taking, and in the extension of these skills in responding on the basis of false belief and deception. The findings are not only of developmental significance, but the training employed in various studies has also highlighted the potential for establishing these critical deictic relations in populations for whom they are found to be absent or deficient.

Deictic Relations, Social Anhedonia, and Schizophrenia

Several studies have examined the potential role of deictic relations in understanding the self, and the findings from these are summarized briefly below. For example, Villatte, Monestès, McHugh, Freixa i Baqué and Loas (2008) compared performances on a 62-trial version of the deictic relations protocol with performances on standard ToM tasks in a non-clinical sample of 60 college students who had scored high or low on a scale of social anhedonia. The results showed a significant correlation on outcomes for these two types of tasks. Although all participants produced better accuracies on simple vs. reversed relations overall, the low-anhedonia group performed equally well on I-YOU reversals as on simple I-YOU relations, while the high-anhedonia group produced relatively weaker performances on the I-YOU reversals. Furthermore, the high-anhedonia group failed to score above chance on both types of double reversal trials, and produced significantly more errors on the I-YOU/ HERE-THERE double reversals than the low-anhedonia group.

A subsequent study by Villatte, Monestès, McHugh, Freixa i Baqué, and Loas (2010a) replicated the previous research with 15 participants with a diagnosis of schizophrenia (aged 22-53 years) and with 15 non-clinical individuals (20-63 years). The results here indicated a considerable pattern of differences between the two groups. 1. The clinical sample produced more errors than control participants on both I-YOU and NOW-THEN reversals, and the latter difference approached significance. 2. The clinical sample also made significantly more errors on all three types of reversed vs. simple matching relations, while the control group did not. 3. The clinical sample produced weaker ToM performances than the control group, and regression analysis indicated that accuracy in reversing deictic relations was a strong predictor of ToM performances for both groups.

Villatte, Monestès, McHugh, Freixa i Baqué, and Loas (2010b) conducted two additional studies. One investigated deictic relations in terms of true and false belief with a non-clinical sample of 30 college students who scored high on the same social anhedonia scale vs. a non-clinical control group. The 48-item protocol comprised four trial-types: self-attribution true-belief (simple I-YOU relations); self-attribution false-belief (simple I-YOU relations with logical not); attribution-to-other true-belief (reversed I-YOU relations); and attribution-to-other false-belief (reversed I-YOU relations with logical not). Consider the self-attribution true-belief trial: "If I put the pencils in the Smarties box and YOU are here, what would YOU think the Smarties box contains?" Participants were presented with two options: Pencils or Smarties. In this example, the correct response is "pencils". The results indicated that the high-anhedonia group produced more errors on both true- and false-belief attributions to other than the control group. Specifically, they produced 75% more errors on all true-belief trials, 300% more errors on the false-belief trials; and more errors overall on attribution-to-other vs. self-attribution trials.

The second study by Villatte et al. (2010b) replicated the previous research with a sample of 15 adults with a diagnosis of schizophrenia, compared with a control group. The results here indicated that the clinical sample produced more errors than controls on the attribution-to-other trials and significantly more errors on the self-attribution false-belief trials. Furthermore, they also produced more errors on self-attribution of false vs. true belief.

Taken together, these findings show a considerable pattern of deficits in responding to deictic relations, in terms of both true and false belief, in samples who present with high social anhedonia and with schizophrenia, relative to controls. These outcomes suggest a possible role for relational perspective-taking in both of these presentations, characterized as having deficits in a sense of self. Indeed, the section below reviews a recent RFT account of this relationship.

Deictic Relations and the Self

The concept of the "three selves" emerged in the earliest literature on Acceptance and Commitment Therapy (ACT, see Hayes, Strosahl, & Wilson 1999). In short, these comprised self as content, self as process, and self as context. Although there has been no further technical account of these concepts, they appear to involve deictic relations (see Foody, Barnes-Holmes, & Barnes-Holmes, 2012).

From an RFT perspective, differences among the three selves depend upon the hypothetical "location" of one's psychological content (i.e., thoughts, feelings, emotions) and the extent to which this content influences behavior. For example, in self as content, your psychological content is located HERE and NOW, but is rigid and attached to your perspective (i.e., I is coordinated with content, hence you are what you think). For ACT, this effect is known as fusion (see Hayes et al., 1999), and is the most difficult place from which to operate because there is no separation between you and your thoughts. In self as process the psychological content is also HERE and NOW, but this content is ever-changing and not attached to one's perspective. That is, although your perspective is not distinct from your content, you recognize that thoughts are ever-changing. In self as context, your content is located THERE and THEN (hence is distinct from your perspective), so that you can recognize that thoughts are only thoughts. When operating at this sense of self, content has little influence over behavior and thus your sense of self is more stable and secure.

Targeting deictic relations may be a possibly useful way of reducing the distressing and ineffective hold psychological content has over behavior and in bolstering a more secure sense of self or perspective. Indeed, some preliminary evidence appears to support this view. Specifically, a recent ACT-based study by Luciano et al. (2011) attempted to establish hierarchical relations between the self and one's thoughts (e.g. "imagine yourself so big as to have room for all of the thoughts you have had today"), and the data suggested that this contributed to improvements in a range of psychological measures for adolescents at high and lower risk of conduct difficulties.

Concluding Comments

The current chapter set out to articulate the RFT account of perspective-taking and to give the reader a flavor of the sizeable expansion of research into the deictic relations in the last decade, since the first full-length RFT book. The many studies reviewed herein not only support the original RFT account and predictions regarding relational perspective-taking, but also show its application to an understanding of typical and atypical development, as well as interventions with clinical populations. In simple terms, the chapter documents ten more years of RFT research that supports the view of perspective-taking as relational responding and suggests the importance of this behavior in many important aspects of living.

References

Barnes-Holmes, Y. (2001). Analysing relational frames: Studying language and cognition in young children. (Unpublished doctoral dissertation). National University of Ireland Maynooth, Ireland.

Barnes-Holmes, Y., Barnes-Holmes, D., & Smeets, P. M. (2004). Establishing relational responding in accordance with opposite as generalized operant behavior in young children. *International Journal of Psychology and Psychological Therapy*, 4(3), 559-586.

Barnes-Holmes, Y., Barnes-Holmes, D., Smeets, P. M., Strand, P., & Friman, P. (2004). Establishing relational responding in accordance with more-than and less-than as generalized operant behavior in young children. *International Journal of Psychology and Psychological Therapy*, 4, 531-558.

Baron-Cohen, S., Tager-Flusberg, H., & Cohen, D. J. (2000). *Understanding other minds: Perspectives from developmental cognitive neuroscience.* Oxford: Oxford University Press.

Berens, N. M., & Hayes, S. C. (2007). Arbitrary applicable comparative relations: Experimental evidence for relational operants. *Journal of Applied Behavior Analysis, 40,* 45-71.

Brüne, M. (2005). "Theory of Mind" in schizophrenia: A review of the literature. *Schizophrenia Bulletin, 31*(1), 21-42.

Davlin, N. L., Rehfeldt, R. A., & Lovett, S. (2011). A relational frame theory approach to understanding perspective-taking using children's stories in typically developing children. *European Journal of Behavior Analysis, 12*(2), 403-430.

Foody, M., Barnes-Holmes, Y., & Barnes-Holmes, D. (2012). The role of self in Acceptance and Commitment Therapy (ACT). In L. McHugh and I. Stewart (Eds.), *The self and perspective taking: Theory and research from contextual behavioral science and applied approaches* (pp. 125-142). Oakland, CA: New Harbinger.

Frith, U., Happé, F., & Siddons, F. (1994). Autism and theory of mind in everyday life. *Social Development, 2,* 108-124.

Gore, N. J., Barnes-Holmes, Y., & Murphy, G. (2010). The relationship between intellectual functioning and relational perspective-taking. *International Journal of Psychology and Psychological Therapy, 10*(1), 1-17.

Gorham, M., Barnes-Holmes, Y., Barnes-Holmes, D., & Berens, N. (2009). Derived comparative and transitive relations in young children with and without autism. *The Psychological Record, 59,* 221-246.

Griffee, K., & Dougher, M. J. (2002). Contextual control of stimulus generalization and stimulus equivalence in hierarchical categorization. *Journal of the Experimental Analysis of Behavior, 78,* 433-447.

Hayes, S. C., Fox, E., Gifford, E., Wilson, K., Barnes-Holmes, D., & Healy, O. (2001). Derived relational responding as learned behavior. In S.C. Hayes, S. Barnes-Holmes, & B. Roche (Eds.), *Relational frame theory: A post-Skinnerian account of human language and cognition.* (pp. 21-50). New York: Plenum.

Hayes, S. C., Strosahl, K. D., & Wilson, K. G. (1999). *Acceptance and Commitment Therapy: An experiential approach to behavior change.* New York: Guilford Press.

Hayes, S., & Wilson, K. G. (1996). Criticisms of relational frame theory: Implications for a behavior analytic account of derived stimulus relations. *The Psychological Record, 46,* 221-236.

Heagle, A., & Rehfeldt, R. A. (2006). Teaching perspective-taking skills to typically developing children through derived relational responding. *The Journal of Intensive Early Behavioral Intervention, 3,* 8-34.

Howlin, P., Baron-Cohen, S., & Hadwin, J. (1999). *Teaching children with autism to mind-read: A practical guide.* Chichester, England: Wiley.

Kohlberg, L. (1976). Moral stages and moralization: The cognitive-developmental approach. In T. E. Lickona (Ed.), *Moral development and behaviour: theory, research, and social issues* (pp.31-55). New York: Holt, Rinehart & Winston.

Leslie, A. M. (1987). Pretense and representation: The origins of "Theory of Mind". *Psychological Review, 94*(4), 412-426.

Luciano, C., Gómez-Becerra, I., & Rodríguez-Valverde, M. (2007). The role of multiple-exemplar training and naming in establishing derived equivalence in an infant. *Journal of the Experimental Analysis of Behavior, 87,* 349-365.

Luciano, C., Ruiz, F. J., Vizcaíno Torres, R. M., Sánchez Martín V., Gutiérrez Martínez, O., & López-López, J. C. (2011). A relational frame analysis of defusion in Acceptance and Commitment Therapy: A preliminary and quasi-experimental study with at-risk adolescents. *International Journal of Psychology and Psychological Therapy, 11*(2), 165-182.

McHugh, L., Barnes-Holmes, Y., & Barnes-Holmes, D. (2004a). Perspective-taking as relational responding: A developmental profile. *The Psychological Record, 54,* 115-144.

McHugh, L., Barnes-Holmes, Y., & Barnes-Holmes, D. (2004b). A relational frame account of the development of complex cognitive phenomena: Perspective-taking, false belief understanding, and deception. *International Journal of Psychology and Psychological Therapy, 4*(2), 303-324.

McHugh, L., Barnes-Holmes, Y., Barnes-Holmes, D., & Stewart, I. (2006). False belief as generalised operant behavior. *The Psychological Record, 56,* 341-364.

McHugh, L., Barnes-Holmes, Y., Barnes-Holmes, D., Stewart, I., & Dymond, S. (2007). Deictic relational complexity and the development of deception. *Psychological Record, 57*(4), 517-531.

McHugh, L., Barnes-Holmes, Y., O'Hora, D., & Barnes-Holmes, D. (2004). Perspective-taking: A relational frame analysis. *Experimental Analysis of Human Behaviour Bulletin, 22,* 4-10.

O'Connor, J., Rafferty, A., Barnes-Holmes, D., & Barnes-Holmes, Y. (2009). The role of verbal behavior, stimulus nameability and familiarity on the equivalence performances of autistic and normally developing children. *The Psychological Record, 59*(10), 53-74.

O'Hora, D., Pelaez, M., & Barnes-Holmes, D. (2005). Derived relational responding and performances on verbal sub-tests of the WAIS-III. *The Psychological Record, 55,* 115-175.

Rehfeldt, R. A., Dillen, J. E., Ziomek, M. M., & Kowalchuk, R. K. (2007). Assessing relational learning deficits in perspective-taking in children with high-functioning autism spectrum disorder. *The Psychological Record, 57,* 23-47.

Tager-Flusberg, H., & Anderson, M. (1991). The development of contingent discourse ability in autistic children. *Journal of Child Psychology and Psychiatry, 32,* 1123-1134.

Villatte, M., Monestès, J-L., McHugh, L., Freixa i Baqué, E., & Loas, G. (2008). Assessing deictic relational responding in social anhedonia: A functional approach to the development of Theory of Mind impairments. *International Journal of Behavioral Consultation and Therapy, 4*(4), 360-373.

Villatte, M., Monestès, J-L., McHugh, L., Freixa i Baqué, E., & Loas, G. (2010a). Assessing perspective taking in schizophrenia using Relational Frame Theory. *The Psychological Record, 60,* 413-436.

Villatte, M., Monestès, J-L., McHugh, L., Freixa i Baqué, E., & Loas, G. (2010b). Adopting the perspective of another in belief attribution: Contribution of Relational Frame Theory to the understanding of impairments in schizophrenia. *Journal of Behavior Therapy and Experimental Psychiatry, 41,* 125-134.

Weil, T. M., Hayes, S. C., & Capurro, P. (2011). Establishing a deictic relational repertoire in young children. *The Psychological Record, 61,* 371-390.

PART III

Applications of Relational Frame Theory

CHAPTER SEVEN

Advances in Language Interventions Based on Relational Frame Theory for Individuals with Developmental Disorders

Clarissa S. Barnes

Ruth Anne Rehfeldt

Southern Illinois University, Carbondale

Traditional behavior analytic instructional programs have typically used discrete trial teaching (DTT) to establish language repertoires in individuals with developmental disorders. Such programs may focus on teaching individuals to respond as listeners, by selecting items from an array, and as speakers, by requesting, labeling, and commenting. A discrete trial is a short unit of instruction, the onset of which is typically marked by an instruction given by the instructor. Correct responses are prompted by the instructor with prompts being faded as the learner acquires the skill. Correct responses are reinforced and error correction is provided contingent on incorrect responding (Smith, 2001). Discrete trial teaching has been demonstrated to be an effective instructional method for individuals with developmental disorders, particularly if delivered at the intensity of 40 hours a week (Lovaas, 1987). These and other results have resulted in the frequent use of DTT as an instructional method in early intervention programs with individuals with language delays. However, skills taught using DTT may not be durable over time, stimuli, settings, and individuals (Schreibman, 2000). DTT is a highly structured form of teaching, with each response evoked by particular stimulus conditions in contrived settings with limited or no distractions (Smith, 2001). These highly structured instructional methods have been criticized for not readily promoting generative responding (Alessi, 1987). That is, traditional DTT instructional methods may be described as teaching specific responses rather than generalized repertoires.

The research program on derived stimulus relations may provide a curricular framework for producing emergent or generative language skills in such a way that reduces some of these historical concerns regarding DTT. In addition, instructional protocols based on this framework may prove to be economic and efficient; in other words, a number of novel or untaught skills may emerge after only a few are explicitly taught, an apparent improvement upon the sometimes tedious and repetitive nature of DTT. The purpose of the current chapter is to outline the history of research conducted to date inspired by Relational Frame Theory that has promoted emergent or generative language in individuals with developmental disorders. We will focus upon both basic and more complex language repertoires.

Relational Frame Theory Approach to Language Instruction

According to Relational Frame Theory (RFT), relational responding, or relating one stimulus in terms of another, is viewed as overarching operant behavior, or a functional class of behaviors (Hayes, Barnes-Holmes, & Roche, 2001; Healy, Barnes-Holmes, & Smeets, 2000; Stewart & Roche, this volume). RFT works under the assumption that if an individual is taught the relations "A is lighter than B, and A is heavier than C," he or she will show the emergence of the relations "B is heavier than A, C is lighter than A, B is heavier than C, and C is lighter than B," in the absence of explicit instruction (see Figure 1).

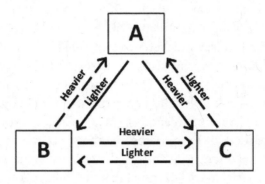

Figure 1. Taught relations (solid lines) and derived relations (dotted lines) for three stimuli participating in a frame of comparison.

In RFT terminology, the derived B-A and C-A relations indicate mutual entailment, and the derived B-C and C-B derived relations demonstrate combinatorial entailment. RFT also predicts the transformation of functions among mutually and combinatorially entailed stimuli. Transformation of stimulus functions occurs when the psychological functions of one stimulus that are relevant in the relational context alter the function of another stimulus in the relation (Hayes et al., 2001). In the present example if the individual had been taught that he or she could lift stimulus A, he or she would derive that he or she could also lift stimulus C, based on the relations that C is lighter than A and A is heavier than C. RFT contends that these emergent untaught skills would be particularly likely if the individual had experienced a history of reinforcement for responding in accordance with multiple exemplars (hereafter referred to as multiple exemplar instruction, or MEI), in which all of the emergent skills were directly taught with other instructional stimuli in similar contexts.

A study by Berens and Hayes (2007) demonstrated the role of MEI in establishing frames of comparison with typically developing children, thus providing support for relational responding as an overarching operant class. Four typically developing 3-4 year old children partici-pated in the study. Three sets of 3 paper coins depicting arbitrary pic-tures were used as stimuli. Each session began with the experimenter telling the participant that they would be playing a game and that the participant should pick whichever picture would buy him or her the most candy. Following these instructions the experimenter placed between two and three pictures in front of the participant and labeled the relation between the stimuli (e.g. "This one [point to one of the stimuli] is more than this one [pointing to another stimulus]"). The experimenter then asked the participant to select the one that he or she would use to buy candy. It is important to note that the assigned value of each picture changed from trial to trial to ensure that the children were responding in accordance with the relational context (Crel) rather than their own experimental history with the pictures. Training consisted of five specific trial types: more than trials with two stimuli (e.g., A>B), less than trials with two stimuli (e.g., A<B), more than trials with three stimuli (e.g., A>B>C), less than trials with three stimuli (e.g., C<B<A), and mixed trials with three stimuli (e.g., A>B<C). Nonarbitrary remedial training sessions were also conducted intermittently. The test probes conducted throughout the study indicated that the children acquired the target response (selecting the picture that is worth more) over the course of training, but that performance did not improve during extended base-lines, helping to rule out alternative explanations (e.g., maturation). Additionally, an analysis of responding during test probe conditions indi-cated that the specific trial types targeted in each training session were the trial types that accounted for the improvement in baseline respond-ing (Berens & Hayes, 2007). For example, the participants' performance on less than trials with two stimuli (e.g., A<B) did not improve during test probes until after training of less than trials with two stimuli. Overall, this experiment provides evidence for the role of MEI in promoting rela-tional responding as a generalized overarching operant class.

Many typically developing children learn relational frames without any formal or specialized language intervention (Greer, Stolfi, Chavez-Brown, & Rivera-Valdes, 2005). Some of the earliest relations, the object-name/name-object frames of coordination, are established via daily interactions between caregivers, children, and the physical environment. For example, as shown in Figure 2, a child may orient toward a ball as the caregiver says, "ball." The caregiver may differentially reinforce the child's echoing of the word "ball". The caregiver may also instruct the child to "look at the ball" and differentially reinforce the child's orienting toward the ball. This sort of MEI for identifying or orienting toward and labeling objects in the environment may result in the child's ability to show one form of responding (i.e., labeling or identifying) with novel stimuli after only being explicitly taught the other skill (i.e., labeling or identifying). When this repertoire of bidirectional responding is not established via naturalistic environmental interactions, more structured intervention may be necessary (Greer et al., 2005). Research has shown that such interventions have been useful in promoting both rudimentary and more complex verbal skills.

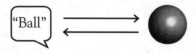

Figure 2. The name-object and object-name relations that many typically developing children learn via natural multiple exemplar instruction with caregivers.

Establishing Basic Verbal Operants

A potentially effective intervention strategy may be to incorporate Skinner's (1957) analysis of verbal behavior with RFT. Skinner's analysis of verbal behavior, which categorizes verbal operants on the basis of each operant's controlling antecedents and consequents, is currently the theoretical framework for a popular curricular model for children with autism (i.e., Sundberg, 2008). Despite the popularity of this curricular approach, RFT theorists would argue that Skinner's analysis has not produced a productive research program on such diverse topics relevant to language instruction as language generativity, problem-solving, and logic (see, for example, Dymond & Alonso-Alvarez, 2010). Skinner's analysis has also been criticized for failing to account for the behavior of the listener (Hayes & Hayes, 1989), and consequently failing to adequately distinguish verbal behavior from other forms of social behavior (Parrott, 1986). Barnes-Holmes, Barnes-Holmes, and Cullinan (2000), however, propose that a synthesis between Skinner's (1957) account and that of RFT may provide a useful framework for continued research on verbal behavior. These authors proposed that a "truly verbal" operant is an operant that emerges on the basis of a history of reinforcement for arbitrarily applicable relational responding, thus distinguishing verbal behavior from other types of social behavior (Barnes-Holmes et al., 2000). The synthesis proposed by these authors has inspired a number of studies related to language intervention.

The mand, defined by Skinner (1957), is a verbal operant that is under the control of motivating operations and maintained by specific reinforcement. Michael (1988 & 1993) discussed the two effects that establishing or motivating (Laraway, Snycerski, Michael, & Poling, 2003) operations have on organisms — a reinforcer establishing effect and an evocative effect. The reinforcer establishing effect momentarily increases the effectiveness of a particular stimulus or event as a reinforcer, whereas the evocative effect momentarily increases the frequency of behaviors previously reinforced by that particular stimulus or event (Michael, 1993). The capturing and contriving of MOs are crucial when developing an instructional protocol to teach mands. Capitalizing on MOs can be done when an individual indicates that a particular MO is in place, whereas contriving MOs requires teachers or caregivers to design situations in which the learner needs an item to complete a task or gain access to another item (Michael, 1988; Michael, 1993). The interrupted chain procedure is a frequently used procedure for establishing mands under the control of contrived MOs. In this procedure, an item necessary to complete a chained task is removed until the learner mands for the item (Hall & Sundberg, 1987).

Derived Mands

A derived mand, as described by RFT, is a mand that participates in a bidirectional, or multidirectional, relational frame (Barnes-Holmes et al., 2000). These authors provide the example of a child who has learned to vocally request "toy" when the child wants a toy car. Because of the participation of the car in a frame of coordination with a doll, the child is also likely to request "toy" when he or she similarly wants the doll, in the absence of direct instruction. Rehfeldt and Root (2005) demonstrated that a history of relational responding via conditional discrimination instruction was sufficient for establishing derived mands according to the Barnes-Holmes et al. (2000) definition. In this case, derived mands consisted of exchanging text for access to corresponding preferred items. Three adults with severe intellectual disabilities were taught to request preferred items by exchanging pictures for those items. Once this initial mand instruction was complete, pretests were conducted. Pretest scores for all participants indicated that none of the participants named pictures or read words corresponding to their preferred items, matched words to pictures or vice versa, or exchanged text for access to preferred items. Following pretest probes, conditional discrimination teaching was introduced, which consisted of teaching participants to match dictated names for preferred items to the corresponding pictures and printed words. Following mastery of the conditional discriminations, the participants matched the text to pictures, pictures to text, and demonstrated derived manding by using the text to request preferred items (Rehfeldt & Root, 2005). Two of the three participants also named the pictures and read the text following conditional discrimination instruction. It should be noted that these mands can be considered impure according to Skinner's (1957) definition in the sense that the preferred items were within the individuals' view, such that the requests were not exclusively under the control of MOs. However, according to Barnes-Holmes et al., the mands could be considered truly verbal considering that they emerged from an instructional history of relating pictures of preferred items to their corresponding text and dictated names.

Similar results have been demonstrated using the interrupted chain procedure, in which the occurrence of derived mands exclusively under the control of MOs was evaluated. Rosales and Rehfeldt (2007) taught two individuals with severe intellectual disabilities to request items needed to complete chained tasks (listening to a portable CD player and making Kool-Aid) by exchanging pictures of the necessary items for the items themselves. Once mands were mastered, conditional discrimination instruction was introduced, in which participants were taught to conditionally relate dictated names of items needed to complete the chained tasks to their corresponding pictures and text, as shown in Figure 3.

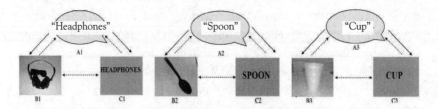

Figure 3. Relations taught (solid lines) and derived (dotted lines) during conditional discrimination training. From "Contriving Transitive Conditioned Establishing Operations to Establish Derived Manding Skills in Adults with Severe Developmental Disabilities," by R. Rosales and R. A. Rehfeldt, 2007, *Journal of Applied Behavior Analysis, 40,* p. 112. Copyright 2007 by Ruth Anne Rehfeldt. Adapted with permission.

Following teaching, both participants matched text to pictures; matched pictures to text; named the pictures, read the text, and importantly, exchanged text for access to the items needed to complete the chains (see Figure 4). Results indicated that the derived mands did not maintain at one-month follow-up, but did remain higher than pretest levels for both participants.

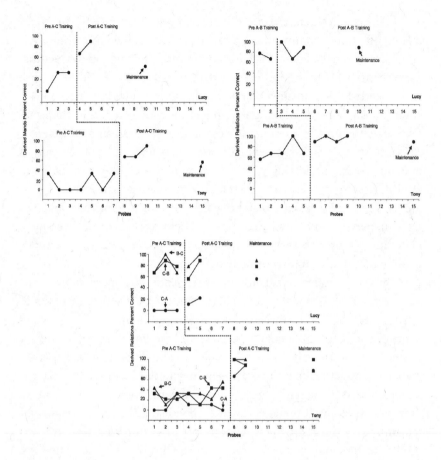

Figure 4. Percentage of correct derived mands after dictated name-text conditional discrimination training (top left panel); percent correct picture naming following dictated name-picture conditional discrimination training (top right panel); percent correct reading text, picture-text matching, and text-picture matching after dictated name-text conditional discrimination training (bottom panel). Each data point indicates a nine-trial block. From "Contriving Transitive Conditioned Establishing Operations to Establish Derived Manding Skills in Adults with Severe Developmental Disabilities," by R. Rosales and R. A. Rehfeldt, 2007, *Journal of Applied Behavior Analysis, 40,* p. 112. Copyright 2007 by Ruth Anne Rehfeldt. Reprinted with permission.

These studies demonstrate the applied relevance of RFT in establishing functional communication repertoires for individuals with developmental disorders. Following basic conditional discrimination teaching, participants in both studies used text to mand for preferred items and items needed to complete a chained task, skills particularly relevant for adult participants for whom communicating with text rather than pictures may be more appropriate in community settings. In addition, participants showed the emergence of picture to text and text to picture matching skills, indicating that they comprehended the text (Sidman, 1971). These studies also demonstrate the generative nature of RFT protocols. Teaching two skills, matching dictated names to pictures and dictated names to text, resulted in the participants acquiring five skills: matching pictures to text, matching text to pictures, tacting pictures, reading text, and manding using text.

Derived manding in accordance with relations of more and less has also been established with individuals diagnosed with autism (Murphy & Barnes-Holmes, 2009). In Murphy and Barnes-Holmes (2009) participants participated in a board game where the goal was to accrue six smiley faces on the game board in order to win. This game was used to contrive motivating operations for manding for more or less. The participants manded by exchanging experimental cards with arbitrary symbols drawn on them with the experimenter. Following the pre-experimental teaching with the board game, conditional discrimination teaching was used to teach more- and less-than relations with the A1 (more) and A2 (less) stimuli. Next, conditional discrimination teaching was conducted to establish frames of coordination between A1-B1 and B1-C1 as well as A2-B2 and B2-C2. Following the conditional discrimination instruction, derived mand probes were conducted in which the individuals were presented with the smiley face board games and the C1 (more) and C2 (less) stimuli in order to mand for the addition or removal of smiley faces in order to win the game. After probing for derived mands, conditional discrimination teaching was conducted to reverse the relations. That is, B1-C2 was taught in a frame of coordination and B2-C1 was taught in a frame of coordination. Next, another test probe was conducted to determine if derived mands would emerge for more (C2) and less (C1). Baseline relations were then retaught (B1-C1 and B2-C2) and a final test, identical to the initial test for derived relations, was administered. Results indicated that all four participants showed the emergence of derived more-less mands following conditional discrimination teaching. Reversed mands were also derived, and all three participants responded correctly to the baseline mands following additional conditional discrimination teaching. The results of this study suggest that flexible mands (indicated by the ability to reverse the mands) can be established via conditional discrimination instruction with clinical populations. This is important when considering the function of a complex mand repertoire in the natural environment, where it is important that one's mand repertoire is not too narrow to be functional.

Derived Tacts

The tact is a verbal operant often referred to in layman's terms as labeling. The tact was defined by Skinner (1957) as a verbal operant that is under the control of a nonverbal stimulus (e.g., the object specified in the tact) and maintained by nonspecific reinforcement. Barnes-Holmes et al. (2000) defined a derived tact as a tact that participates in a bidirectional, or multidirectional, relational frame; in other words, a tact that is not explicitly taught but rather emerges from a history of reinforcement for relational responding. Programming for the emergence of derived tacts may have an important practical benefit. Namely, many stimuli have multiple names in conventional language, and by programming for the emergence of untaught tacts, an individual need not be directly instructed on each and every hierarchical name for every stimulus item.

Conditional discrimination instruction has been shown to be effective in producing untaught tacts in adults with severe intellectual disabilities (Halvey & Rehfeldt, 2005). These authors first identified preferred items in three adults with severe intellectual disabilities. Three categories of preferred items were identified for each participant. Within each category three specific stimuli were identified and assigned alphanumeric labels (see Table 1).

Table 1.

Preferred Categories and Stimuli for Each Participant Described in
Halvey and Rehfeldt (2005) with Alphanumeric Labels

Participants	Category	Stimuli		
Josh, Ray, and Jayne	Music	CD (A1)	Walkman (B1)	Portable Radio (C1)
Josh and Ray	Money	Dollar Bill (A2)	Loose Change (B2)	Wallet (C2)
Josh, Ray, and Jayne	Fruit	Piece of Pear (A3)	Piece of Apple (B3)	Piece of Banana (C3)
Jayne	Movie	Portable TV (A3)	Videotape (B3)	Portable VCR (C3)

After identifying preferred items, pretest probes were conducted to evaluate participants' manding and tacting of items using their category names, as well as their abilities to conditionally relate photographs of stimuli belonging to the same category. Following pretest probes a graduated time-delay was used to teach each participant to request their preferred items, the "B" stimuli, using the corresponding category names. Visual-visual conditional discrimination instruction using items from the different categories was then conducted. Figure 5 shows an example of the relations taught during conditional discrimination teaching, as well as the relations tested during pre- and posttest probes. Specifically, the experimenters taught the participants to conditionally relate pictures of A stimuli to the B stimuli and the A stimuli to the C stimuli within each category.

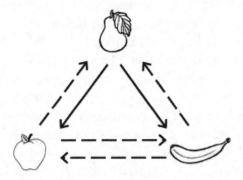

Figure 5. Conditional discriminations taught (solid lines) and derived (dotted lines) for the fruit category in Halvey and Rehfeldt (2005).

Following mand and conditional discrimination teaching, posttest probes (identical to pretest probes) were conducted with each participant. Results suggested that mand training and a history of reinforced conditional discriminations can facilitate not only derived mands in learners with disabilities, as previous research has indicated, but also derived tacts. In other words, participants tacted stimuli using their category names without explicit instruction, but merely on the basis of a history of conditionally relating those stimuli to other categorically related stimuli. A flexible tact repertoire may permit individuals to communicate fluidly in a variety of contexts and situations. Programming for the emergence of untaught tacts may also decrease the time and resources necessary to individually teach each target response.

Derived Intraverbals

The intraverbal, according to Skinner's (1957) analysis, is a verbal operant that is under the control of a verbal stimulus that does not have point-to-point correspondence with the response and is maintained by nonspecific reinforcement. In educational curricula the intraverbal is typically thought of as conversation. Barnes-Holmes et al. (2000) define a derived intraverbal as an intraverbal that participates in a bidirectional, or multidirectional, relational frame. Many individuals diagnosed with autism or other developmental disabilities often struggle or fail to acquire functional intraverbal repertoires (Sundberg & Sundberg, 2011). Often, specific intraverbals targeted for intervention are taught directly using tedious transfer of stimulus control procedures (Grannan & Rehfeldt, 2012), which may result in intraverbals that are rote and can only be evoked by specific stimulus conditions. Utilizing protocols that facilitate derived intraverbal responding may result in the establishment of intraverbal repertoires that are verbal, functional, and flexible.

MEI has been demonstrated to be effective in establishing derived intraverbals. Perez-Gonzalez, Garcia-Asenjo, Williams, and Carnerero (2007) used MEI to establish derived intraverbal antonyms with children diagnosed with pervasive developmental disorders. Specifically, the researchers taught participants to respond to intraverbals that included antonyms. For example, given the discriminative stimulus, "Tell me the opposite of up," the child was taught to respond "down," and in the presence of the discriminative stimulus, "Tell me the opposite of down," the child was expected to show the emergence of the untaught response, "up". Only one intraverbal response was taught to the participants in the study (e.g., only the "down" response). Following teaching, the reverse intraverbal was probed (e.g., the "up" response). If participants did not respond correctly to the reverse intraverbals, the reverse intraverbals for that set were directly taught. Following direct teaching of both the original and reverse intraverbals (e.g., the "down" and "up" response), a new set was introduced and the teaching test probe pattern was repeated. Prior to MEI neither participant showed the reverse intraverbals. Following MEI both participants correctly responded to untaught reverse intraverbal probes. These results support the use of MEI to expand intraverbal repertoires of individuals with developmental disabilities. While

MEI was effective in establishing emergent derived intraverbals, the concept of opposites was not evaluated in this study. That is, it is unclear if the relational cue "opposite" had any effect on participants' responses. Establishing a frame of opposition was not necessary for the participants to show the reverse intraverbals.

Protocols utilizing tangible stimuli have also been demonstrated to be effective in establishing untaught intraverbals. Grannan and Rehfeldt (2012) evaluated the effects of tact and match-to-sample instruction on categorization intraverbals with two children diagnosed with autism. Initially the participants were taught to tact items from four different categories (e.g., vehicles, clothing, body parts, and musical instruments). Following simple tact teaching (e.g., "What is it?" "A car.") the participants were taught to tact the categories for each item (e.g. "What is a car?" "A vehicle."). After category tact instruction a match-to-sample protocol was used to teach the participants to match pictures of the items by category. After instruction on both category tacting and match-to-sample, both participants demonstrated derived intraverbals. That is, without direct teaching, both participants listed items belonging to particular categories when asked to do so (Grannan & Rehfeldt, see Figure 6).

A flexible intraverbal repertoire is beneficial to individuals in a variety of academic and social settings. Intraverbals are frequently targeted in small group and individual instructional arrangements, and many of one's daily social interactions consist of intraverbals. Without a repertoire of derived intraverbals, individuals with autism and other disorders may be limited to emitting rote responses that may not be functional in multiple contexts. If generative procedures such as those reported by Grannan and Rehfeldt are implemented, an individual may be able to respond to a variety of questions with a variety of responses (e.g., May, Hawkins, & Dymond, in press).

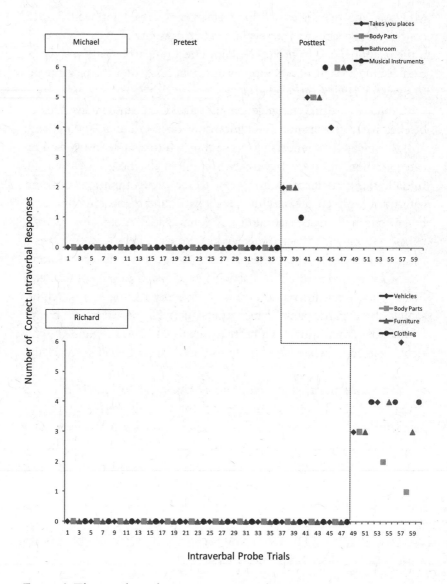

Figure 6. The number of correct responses for intraverbal categorization during pre- and posttest probes for both participants. From "Emergent Intraverbal Responses via Tact and Match-to-Sample Instruction," by L. Grannan and R. A. Rehfeldt, 2012, *Journal of Applied Behavior Analysis*, 45. Reprinted with permission.

Second Language Learning

MEI has been shown to be effective in facilitating derived tacts in preschool children learning English as a second language. Rosales, Rehfeldt, and Lovett (2011) used MEI to facilitate the emergence of derived tacts following listener teaching with four typically developing Spanish speaking preschoolers learning English as a second language. The children were taught to respond as listeners when the experimenter presented dictated names in English for a number of everyday items. If the children did not tact the stimuli following listener instruction, MEI with a novel assortment of stimuli was introduced, in which the participants were taught both listener and tact responses for the stimuli. MEI was shown to be effective in producing untaught naming or tacting of the original instructional stimuli following listener instruction alone (Rosales et al., 2011).

Operant learning alone may not always account for the acquisition of early language skills. Rather, respondent processes may also play a role. During respondent-type teaching, vocal names and their corresponding visual stimuli are presented simultaneously or in close succession, with no specific contingency in place for the child to respond (see Leader & Barnes-Holmes, 2001; Leader, Barnes-Holmes, & Smeets 2000). For example, while feeding a child, a caregiver may say, "Yum! applesauce!" and deliver applesauce to the child. Over time, a typically developing child will likely learn that the spoken word "applesauce" and the actual bowl of applesauce are related in a frame of coordination despite the fact that no response was ever differentially reinforced. Such natural pairings of names and their referents are no doubt abundant in language-rich environments, underscoring the utility that such pairing procedures may have in instructional programs. Instructional protocols based on respondent-type procedures may actually be more beneficial than conditional discrimination tasks because they may be less demanding and tedious for the learner, and can be implemented in a cost effective manner because multiple learners may observe the teaching simultaneously (Rosales, Rehfeldt, & Huffman, 2012). Rosales et al. (2012) used a respondent-type teaching protocol to teach basic vocabulary to children who were learning English as a second language. Participants were three preschool students who spoke Spanish as their first language. The experimenter first conducted a respondent-type teaching procedure where the item was shown to the participants one-by-one on each trial, and the experimenter simultaneously presented the name of the item. Following respondent-type teaching, tact and listener responding probes were conducted to determine if the teaching had facilitated the emergence of untaught object-name and name-object relations. If participants did not perform to criteria following respondent-type teaching, MEI was implemented. During MEI additional sets of stimuli were used and listener and tact relations were explicitly taught. Results from this study highlighted the use of respondent-type teaching for establishing derived listener responses; however, all three participants required MEI on at least one set of stimuli before showing the emergence of untaught tacts (Rosales et al., 2012).

Together these results suggest the powerful role that a learning history with multiple exemplars of object-name and name-object relations may play in promoting the emergence of untaught, generalizable language skills. These results also point to the important place that MEI may occupy in intervention programs for persons with severe language delays.

Deictic Frames and Perspective-Taking

Deictic relations are among the most complex of all relations studied by RFT researchers. In deictic relations, the speaker changes perspective between I-YOU, HERE-THERE, NOW-THEN, or combinations of the three types of relations (Barnes-Holmes, Hayes, & Dymond, 2001). For example, a child who is asked the question "Where were you yesterday?" is learning the coordination between him- or her- self and the "you" in the question, as well as the coordination between him- or her- self and a past location at another time (e.g., there, yesterday) (McHugh, Barnes-Holmes, & Barnes-Holmes 2009). Although responding in accordance with deictic frames often occurs in combination of the different types of relations, researchers have made functional distinctions between the relations for purposes of investigation (McHugh, et al. 2009). Experimentally, deictic relations have been divided into three levels of relational complexity, including simple, reversed, and double reversed (Villatte, Monestes, McHugh, Freixa i Baque, & Loas, 2010). A simple relation is a relation that requires the individual to change perspective according to a single frame (McHugh, Barnes-Holmes, & Barnes-Holmes, 2004). For example, a simple I-YOU relation may be as follows: "I am eating a peanut butter sandwich and you are eating a turkey sandwich. Which sandwich am I eating? Which sandwich are you eating?" A reversed relation requires reversal of a simple relation (e.g., "I am sitting here on the couch and you are sitting there on the chair. If here was there and there was here: where would you be sitting? Where would I be sitting?" McHugh et al., 2004). Double reversed relations require the individual to reverse two simple relations (e.g., "Yesterday I was there watching a movie, today I am here reading a book. If here was there and there was here and if now was then and then was now: Where would I be then? Where would I be now?"; McHugh et al., 2004). RFT researchers have proposed that deictic relations underlie the complex social skill known as perspective-taking, a deficit of which is commonly observed in individuals with a variety of clinical diagnoses. RFT proposes that when two people are interacting, the speaker will always be speaking from I/HERE/NOW to the listener, who is YOU/THERE/THEN, but over the course of various interactions the speaker will be required to respond to various deictic relations in order to relate to and respond appropriately to the verbal behavior of the listener (Villatte et al., 2010). A perspective-taking repertoire is socially relevant in soliciting and giving advice (e.g., "What would you do if you were me?") and empathy (e.g., "Allison just lost her job; how would I feel if I were her?"). Navigating social situations requires individuals to understand what others may be thinking or feeling, as well as make future predictions about their own behavior or the behavior of others.

Deictic relations may also be able to explain the cognitive concept of Theory of Mind (ToM). Two well-known ToM tasks are the Sally Anne task (Baron-Cohen, Leslie, & Frith, 1985) and the hinting task (Corcoran, Mercer, & Frith, 1995). The Sally Anne task is conducted by the experimenter first telling a participant a story about two girls who are playing together. While they are playing, one of the girls, Sally, puts a ball in a basket and then leaves. While Sally is gone Anne moves the ball from the basket to a box. The participant is then asked where Sally will look for the ball when she returns. In order for the individual to answer correctly, he or she must be able to take Sally's perspective. That is, he or she must reverse the I-YOU relation (Villatte et al., 2010). The hinting task also requires the individual to take another's perspective. The hinting task is conducted by reading the individual a short passage describing an everyday interaction between two people (see Villatte, Monestes, McHugh, Freixa i Baque, & Loas, 2008 for more detail). The individual is then asked what one of the characters really means to say. In order to respond correctly, the participant must change perspective to the character (I-YOU), the setting the character is in (HERE-THERE), and when the interaction occurred (NOW-THEN) (Villatte et al., 2010). Research comparing the performance of individuals with schizophrenia to the performance of individuals with no clinical diagnoses showed that the common ToM deficits identified in individuals with schizophrenia correlates with deficits in deictic frame tasks. Specifically, Villatte et al. (2010) showed that performance on perspective-taking tasks predicted accuracy on a hinting task for both individuals with schizophrenia and individuals with no such diagnoses, and participants with schizophrenia did not perform as well as individuals without diagnoses on either task (Villatte et al., 2010).

To establish an RFT foundation for perspective-taking, McHugh et al. (2004) developed a protocol to test simple, reversed and double reversed relations. The protocol consisted of eight simple relations with at least two I-YOU trials, two HERE-THERE trials, and two NOW-THEN trials; 36 reversed relations consisting of eight I-YOU trials, 12 HERE-THERE trials, and 16 NOW-THEN trials; and 18 double reversed relations with six I-YOU/HERE-THERE trials and 12 HERE-THERE/NOW-THEN trials. The protocol was administered to five groups, including adults (18-30 years-old), adolescents (12-14 year-old), children in late childhood (9-11 years-old), children in middle childhood (6-8 years-old) and children in early childhood (3-5 years-old). No feedback for correct or incorrect responses was delivered for participants who completed the protocol. Results indicated that errors decreased as a function of age, with the early childhood group making the most errors, particularly on the reversed and double reversed relations. These findings suggest that the younger children did not yet have the relational repertoires necessary to respond correctly on reversed and double reversed relations trials. These findings were supported in a second study during which the investigators used foil trials to determine if the number of words in the questions for reversed and double reversed relations may have accounted for the increased errors. That is, the investigators used trials that had the same number of words, but did not require relational responding, in order to control for question length with the young children (see Table 2). Children who responded to the foil trials made fewer errors than children who responded to the original trials, indicating that the increased word length was not the cause for the increased number of errors on the reversed and double reversed relations in the original experiment: Increased errors was rather likely due to relational complexity (McHugh et al., 2004).

Table 2.

Examples of a foil trial, a trial that does not require relational responding (top), and an original trial, a trial that does require relational responding (bottom). The bold text indicates the differences in the trials.

Yesterday you were playing soccer, today you are working with me. **If now is now and then was then,** What were you doing then? (Playing soccer) What are you doing now? (Working with you)
Yesterday you were playing soccer, today you are working with me. **If now is then and then was now,** What were you doing then? (Working with you) What are you doing now? (Playing soccer)

Rehfeldt and colleagues have extended the use of the protocol first reported by McHugh et al. (2004). First, Rehfeldt, Dillen, Ziomek, and Kowalchuk (2007) used the same protocol to evaluate differences in perspective-taking between adolescents with high-functioning autism and their age-matched typically developing peers. The authors found that participants with autism made more errors on the task than their typically developing peers, with most of the errors occurring on the reversed relations. In a second experiment, the authors found that the deictic relations were sensitive to reinforcement contingencies and could be readily established using differential reinforcement (Rehfeldt et al., 2007). Davlin, Rehfeldt, and Lovett (2011) developed a more naturalistic deictic frame protocol which could be used over the course of an engaging childhood activity, in this case, reading stories. The authors' protocol required first and second grade children to read stories with the experimenter, over the course of which the experimenter presented a deictic frame protocol requiring participants to change perspective with the character, time, and place occurring in the respective stories. The authors first showed that participants made a number of errors on reversed and double reversed relations, but after receiving instruction with novel books and novel questions, showed the emergence of untaught reversed and double reversed relations with the original test protocol and the original books. These results suggest that, like other relational operants, deictic relations are indeed established via a history of reinforcement for responding in accordance with multiple exemplars. Perspective-taking can thus be regarded as an overarching, generalized response class (Rehfeldt et al., 2007).

Future Directions

Early efforts to use behavior analytic strategies to teach simple and complex language to individuals with developmental disorders often failed to produce flexible, generalizable repertoires. When the natural environment has failed to support the role of multiple exemplars in producing relational abilities, instructional strategies based on RFT can be easily incorporated into other existing curricular framework (see Rehfeldt & Barnes-Holmes, 2009), thus resulting in the emergence of a variety of untaught verbal skills. Future research should elucidate what variables affect the success of naturalistic MEI, and why it is sufficient for some learners but not others.

Many of the academic skills that young children learn in preschool and early elementary include responding in accordance with various relational operants, which are often more complex than relations of equivalence or sameness. For example, understanding relationships of items and numbers (greater than, less than, same as), nonstandard measurements (bigger than, smaller than), sequencing (before, after, now, later), awareness of geographical locations (closer than, farther than), and understanding weather patterns (warmer than, colder than) (Illinois State Board of Education, 2002) are all educational objectives that require the individual to respond in accordance with a variety of different relational frames. Children with disabilities or from language-impoverished environments may benefit greatly from protocols for teaching such skills based on RFT. Much RFT language intervention research conducted to date has focused upon relations of sameness, or equivalence. Future research should focus upon the development of more complex academic interventions involving other frames, and how such interventions can be used to remediate learning deficits in academically challenged populations. Needed also are protocols illustrating how relational repertoires can be established outside the context of structured, laboratory clinical experiences, but rather how parents, caregivers, teachers, and others can utilize MEI more naturalistically to establish relational repertoires in their students and children (see Rehfeldt, 2011).

References

Alessi, G. (1987). Generative strategies and teaching for generalization. *The Analysis of Verbal Behavior, 5*, 15-27.

Barnes-Holmes, D., Barnes-Holmes, Y., & Cullinan, V. (2000). Relational frame theory and Skinner's *Verbal Behavior:* A possible synthesis. *The Behavior Analyst, 23*, 69-84.

Barnes-Holmes, D., Hayes, S. C., & Dymond, S. (2001). Self and self-directed rules. In S. C. Hayes, D. Barnes-Holmes, & B. T. Roche (Eds.), *Relational frame theory: A post-Skinnerian account of language and cognition.*(pp. 21-49). New York: Plenum.

Baron-Cohen, S., Leslie, A. M., & Frith, U. (1985). Does the autistic child have theory of mind? *Cognition, 21*, 37-46.

Berens, N. M., & Hayes, S. C. (2007). Arbitrarily applicable comparative relations: Experimental evidence for a relational operant. *Journal of Applied Behavior Analysis, 40*, 45-71.

Corcoran, R., Mercer, G., & Frith, C. D. (1995). Schizophrenia, symptomology, and social inference: Investigating theory of mind in people with schizophrenia. *Schizophrenia Research, 17*, 5-13.

Davlin, N. L., Rehfeldt, R. A., & Lovett, S. (2011). A relational frame theory approach to understanding perspective-taking using children's stories in typically developing children. *European Journal of Behavior Analysis, 12*, 403-430.

Dymond, S., & Alonso-Álvarez, B. (2010). The selective impact of Skinner's *Verbal Behavior* on empirical research: A reply to Schlinger (2008). *The Psychological Record, 60*, 355–360.

Grannan, L., & Rehfeldt, R. A. (2012). Emergent intraverbal responses via tact and match-to-sample instruction. *Journal of Applied Behavior Analysis, 45*, 601-605.

Greer, R. D., Stolfi, L., Chavez-Brown, M., & Rivera-Valdes, C. (2005). The emergence of the listener to speaker component of naming in children as a function of multiple exemplar instruction. *The Analysis of Verbal Behavior, 21*, 123-134.

Hall, G., & Sundberg, M. L. (1987). Teaching mands by manipulating conditioned establishing operations. *The Analysis of Verbal Behavior, 5,* 41-53.

Halvey, C., & Rehfeldt, R. A. (2005). Expanding vocal requesting repertoires via relational responding in adults with severe developmental disabilities. *The Analysis of Verbal Behavior, 21,* 13-25.

Hayes, S. C., Barnes-Holmes, D., & Roche, B. (Eds.). (2001). *Relational Frame Theory: A post-Skinnerian account of human language and cognition.* New York: Kluwer Academic/Plenum Publishers.

Hayes, S. C., & Hayes, L. J. (1989). The verbal action of the listener as a basis for rule-governance. In Hayes, S. C. (Ed.) *Rule-governed behavior: cognition, contingencies, and instructional control.* (pp. 153-190). New York, NY: Plenum Press.

Healy, O., Barnes-Holmes, D., & Smeets, P. M. (2000). Derived relational responding as generalized operant behavior. *Journal of the Experimental Analysis of Behavior, 74,* 207-227.

Illinois State Board of Education. (2002). *Illinois Early Learning Standards* (No. 504). Springfield, IL.

Laraway, S., Snycerski, S., Michael, J., & Poling, A. (2003). Motivating operations and terms to describe them: Some further refinements. *Journal of Applied Behavior Analysis, 36,* 407-414.

Leader, G., & Barnes-Holmes, D. (2001). Matching to sample and respondent-type teaching as methods for producing equivalence relations: Isolating the critical variable. *The Psychological Record, 51,* 429-444.

Leader, G., Barnes-Holmes, D., & Smeets, P. M. (2000). Establishing equivalence relations using a respondent-type procedure III. *The Psychological Record, 50,* 63-78.

Lovaas, O. I. (1987). Behavioral treatment and normal educational intellectual functioning in young autistic children. *Journal of Consulting and Clinical Psychology, 55,* 3-9.

May, R. J., Hawkins, E., & Dymond, S. (in press). Effects of tact training on emergent intraverbal vocal responses in adolescents with autism. *Journal of Autism and Developmental Disorders.*

McHugh, L., Barnes-Holmes, Y., & Barnes-Holmes, D. (2004). Perspective-taking as relational responding: A development profile. *The Psychological Record, 54,* 115-144.

McHugh, L., Barnes-Holmes, Y., & Barnes-Holmes, D. (2009). Understanding and training perspective taking as relational responding. In R. A. Rehfeldt & Y. Barnes-Holmes (Eds.), *Derived Relational Responding: Applications for Learners with Autism and Other Developmental Disabilities* (pp. 281-287). Oakland, CA: New Harbinger Publications, Inc.

Michael, J. (1988). Establishing operations and the mand. *The Analysis of Verbal Behavior, 6,* 3-9.

Michael, J. (1993). Establishing operations. *The Behavior Analyst, 16,* 191-206.

Murphy, C., & Barnes-Holmes, D. (2009). Derived more-less relational mands in children diagnosed with autism. *Journal of Applied Behavior Analysis, 42,* 253-268.

Parrott, L. J. (1986). On the differences between verbal and social behavior. In P. N. Chase & L. J. Parrott (Eds.), *Psychological aspects of language: The West Virginia Lectures.* Springfield, IL: Thomas.

Perez-Gonzalez, L. A., Garcia-Asenjo, L., Williams, G., & Carnerero, J. J. (2007). Emergence of intraverbal antonyms in children with pervasive developmental disorder. *Journal of Applied Behavior Analysis, 40,* 697-701.

Rehfeldt, R. A. (2011). Toward a technology of derived stimulus relations: An analysis of articles published in *JABA*, 1992-2009. *Journal of Applied Behavior Analysis, 44,* 109-119.

Rehfeldt, R. A., & Barnes-Holmes, Y. (Eds.). (2009). *Derived Relational Responding: Applications for Learners with Autism and Other Developmental Disabilities.* Oakland, CA: New Harbinger Publications, Inc.

Rehfeldt, R. A., & Root, S. L. (2005). Establishing derived requesting skills in adults with severe developmental disabilities. *Journal of Applied Behavior Analysis, 38,* 101-105.

Rehfeldt, R. A., Dillen, J. E., Ziomek, M. M., Kowalchuk, R. K. (2007). Assessing relational learning deficits in perspective taking in children with high-functioning autism spectrum disorders. *The Psychological Record, 57,* 23-47.

Rosales, R., & Rehfeldt, R. A. (2007). Contriving transitive conditioned establishing operations to establish derived manding skills in adults with severe developmental disabilities. *Journal of Applied Behavior Analysis, 40,* 105-121.

Rosales, R., Rehfeldt, R. A., & Huffman, N. (2012). Examining the utility of the stimulus pairing observation procedure with preschool children learning a second language. *Journal of Applied Behavior Analysis, 45,* 173-177.

Rosales, R., Rehfeldt, R. A., & Lovett, S. (2011). An evaluation of multiple exemplar teaching on derived relations in preschool children learning a second language. *The Analysis of Verbal Behavior, 27,* 61-74.

Schreibman, L. (2000). Intensive behavioral/psychoeducational treatments for autism: Research needs and future directions. *Journal of Autism and Developmental Disorders, 30,* 373-378.

Sidman, M. (1971). Reading and auditory-visual equivalences. *Journal of Speech and Hearing Research, 14,* 5-13.

Skinner, B. F. (1957). *Verbal Behavior.* Cambridge, MA: Prentice Hall, Inc.

Smith, T. (2001). Discrete trial training in the treatment of autism. *Focus on Autism and Other Developmental Disabilities, 16,* 86-92.

Sundberg, M. L. (2008). *Verbal Behavior Milestones Assessment and Placement Program.* Concord, CA: AVB Press.

Sundberg, M. L., & Sundberg, C. A. (2011). Intraverbal behavior and verbal conditional discriminations in typically developing children and children with autism. *The Analysis of Verbal Behavior, 27,* 23-43.

Villatte, M., Monestès, J. L., McHugh, L., Freixa i Baqué, E., & Loas, G. (2008). Assessing deictic relational responding in social anhedonia: A functional approach to the development of Theory of Mind impairments. *International Journal of Behavioral Consultation and Therapy, 4*(4), 360-373.

Villatte, M., Monestes, J., McHugh, L., Freixa i Baque, E., & Loas, G. (2010). Assessing perspective taking in schizophrenia using relational frame theory. *The Psychological Record, 60*, 413-436.

CHAPTER EIGHT

Education, Intellectual Development, and Relational Frame Theory

Ian Stewart

National University of Ireland, Galway

Jonathan Tarbox

Center for Autism and Related Disorders

Bryan Roche

National University of Ireland, Maynooth

Denis O'Hora

National University of Ireland, Galway

Introduction

There is now considerable evidence to support the Relational Frame Theory (RFT) position that arbitrarily applicable relational responding or relational framing is the core behavior that characterizes human language and cognition across contexts. The capacity to frame relationally correlates with linguistic and cognitive performance more generally and specifically with measures of both intellectual ability and educational attainment. It makes sense therefore, that, by training relational framing, intellectual performance and educational attainment can be enhanced. The current chapter reviews current research in support of this exciting proposition.

When the first edited volume on RFT was published in 2001, there was already evidence of a link between verbal ability and coordinate framing in the form of stimulus equivalence. However, there was very little work investigating other types of relations and there was no empirical evidence demonstrating that relational framing correlated with intellectual ability as typically measured. In their chapter on education, Barnes-Holmes, Barnes-Holmes and Cullinan (2001) suggested that multiple exemplar training (MET) of relational framing might lead to improved linguistic and intellectual performance. However, there had as yet been no published work demonstrating the efficacy of MET even for improving relational framing itself, let alone showing that training the latter might lead to improvements in cognition and / or intelligence. After an additional decade of RFT research conducted in the meantime, however, there have been impressive advances in all these respects.

Correlations between Relational Framing and Linguistic and Cognitive Abilities

Coordination (2001-2011)

According to RFT, coordinate (sameness) framing is fundamental in language since it underlies linguistic reference. From this perspective, a well-honed ability to respond to and derive frames of coordination is crucial for the acquisition of a broad and well-organized vocabulary. The latter is a basic prerequisite to normal language development, the extent of which is both predicted by and predicts general intelligence (e.g., Smith, Smith, Taylor & Hobby, 2005).

As mentioned above, in 2001 there was already evidence that coordinate framing correlated with verbal ability. For example, research had shown that while humans with even minimal verbal ability readily demonstrated stimulus equivalence, non-humans and non-verbally able humans did not (e.g., Barnes, McCullagh, & Keenan, 1990; Dugdale & Lowe, 2000). Furthermore, empirical effects such as semantic priming had been shown for stimuli in equivalence relations (e.g., Hayes & Bissett, 1998) and similar patterns of neural activity had been found for derived equivalence as for linguistic "processing" (e.g., Dickins, Singh, Roberts, Burns, Downes, Jimmieson & Bentall, 2001).

Over the last decade, further evidence of the correlation between relational framing and linguistic and cognitive abilities has emerged. Many of the relevant studies have continued to feature coordinate framing. For example, Barnes-Holmes et al. (2005 a & b) reported that derived equivalence by typically developing adults both showed semantic priming effects and also produced an event-related potentials (ERPs) signature similar to that for language processing; O'Connor, Rafferty, Barnes-Holmes & Barnes-Holmes (2009) showed that verbal competence of children with ASD (assessed via the Comprehensive Application of Behavior Analysis to Schooling [CABAS; e.g., Greer, 1991] system) correlated with derived equivalence performance; while Moran, Stewart, McElwee & Ming (2010) showed that scores on the communication subscale of the Vineland Assessment of Behavior Scales (VABS) of children with ASD also correlated with the ability to derive equivalence relations.

A number of RFT studies since 2001 have demonstrated a link between derived relational responding and cognitive performance by using derived relational procedures to teach educationally relevant skills. In one such study, Leader and Barnes-Holmes (2001) taught children fraction-decimal equivalence using a respondent-type training procedure and also showed generalization to several new contexts. A series of more recent studies has focused on transfer of functions via equivalence relations and in particular on the transfer of the practically important "mand" response. A mand is a verbal operant under antecedent control of motivating operations that specifies its reinforcer (e.g., the request "cookie" is reinforced by receiving a cookie). A number of studies have demonstrated the efficacy of equivalence procedures in allowing this important skill to transfer rapidly to a variety of new contexts (e.g., Rehfeldt & Root, 2005; Murphy, Barnes-Holmes & Barnes-Holmes, 2005; Murphy & Barnes-Holmes, 2009a, b).

Comparison

Comparison relations are involved "whenever one event is responded to in terms of a quantitative or qualitative relation along a specified dimension with another event" (Hayes, Fox, Gifford, Wilson, Barnes-Holmes & Healy, 2001). Comparison relations are critically important in mathematics and also in our everyday language, which includes diverse sub-types (e.g., faster-slower, older-younger, better-worse etc.). Hence, this frame is particularly important from an educational point of view. A number of studies since 2001 have demonstrated a correlation between comparative relational framing and cognitive performance.

One such study was by Reilly, Whelan and Barnes-Holmes (2005), who investigated the effect of differential training histories on derived comparative relations and found that latencies were significantly lower for those given baseline training exclusively in "more than" relations than for those given either "less than" or a mix of more and less training. This pattern accords with previous studies on relational inferences reported in mainstream cognitive studies. More recently, Munnelly, Dymond and Hinton (2010) used comparative relational framing as a model of the phenomenon of transitive inference and replicated several important effects from the mainstream literature in this domain.

Finally, in a study of direct educational relevance, Murphy and Barnes-Holmes (2010) showed the transformation via comparative relations of derived manding functions in four adolescents with autism and three typically developing children. More and less functions were first established for two arbitrary stimuli and then these were employed to establish a linear network of comparative relations among a further five arbitrary stimuli. Results of subsequent tests showed a derived transformation of functions for several participants.

Temporal Relations

Temporal relational framing involves responding under the control of the contextual cues "Before" and "After" or their functional equivalents. This form of framing, which can be thought of as a specific sub-type of comparative framing, underlies the capacity to coordinate with other individuals and society more generally with regard to the conventionally important dimension of time.

A series of studies have investigated the correlation between temporal relational responding and performance on standard measures of intelligence. O'Hora, Pelaez and Barnes-Holmes (2005) divided college student volunteers into 2 groups on the basis of their performance on a complex relational task that involved responding in accordance with temporal, sameness and distinction relations before subsequently exposing them to the vocabulary, arithmetic, and digit-symbol encoding subtests of the Wechsler Adult Intelligence Scale (WAIS-III). Those successful on the relational task showed superior performance on both the vocabulary and arithmetic subscales in comparison with those who had failed. In a follow-up study, O'Hora, Pelaez, Barnes-Holmes, Rae, Robinson and Chaudhary (2008) compared performance on the same relational task with performance on the entire WAIS-III and showed that successful completion predicted superior performance on both the verbal comprehension and perceptual organization indices.

Recently, O'Toole and Barnes-Holmes (2009) used the Implicit Relational Assessment Procedure (IRAP), a methodology that enables measurement of speed of relational responding, to examine framing in accordance with sameness, distinction and temporal relations and to compare performance with that on the Kaufman Brief Intelligence Test (K-BIT). For each relational task, reaction times were measured first on consistent (in accordance with pre-established verbal relations) trials, and then on inconsistent (against pre-established relations) trials. A difference-score was calculated by subtracting consistent from inconsistent response latencies. The inconsistent trials and the difference-score provided measures of relational flexibility. Results showed that faster responding on the IRAP and smaller difference-scores predicted higher IQ. These findings provided further evidence of the correlation between relational framing and intellectual performance and suggested in particular the importance of relational flexibility as a predictor of intelligent behavior.

Analogical Relations

Analogy is a fundamentally important element of human cognition. It is a critical tool for communication in educational and scientific contexts and is also commonly employed as a metric of intelligent behavior. As such, RFT research into analogy is vital with respect to the provision of comprehensive relational framing based educational intervention.

Barnes, Hegarty and Smeets (1997) provided the first relational frame model of analogy as deriving a relation of sameness between sameness relations ("equivalence-equivalence responding"). This model was demonstrated primarily with adults. Since 2001, that work has been extended with a series of studies comparing analogical framing in different age groups including adults, nine- and five-year olds. Carpentier, Smeets and Barnes-Holmes (2002) showed that whereas the latter two groups could readily show equivalence-equivalence, the five-year olds could not show it unless given specific remediation training. Carpentier, Smeets and Barnes-Holmes (2003) extended these findings by showing that, in contrast with both other age groups, five-year olds were also unable to demonstrate equivalence-equivalence under any circumstances unless first tested for equivalence. Collectively these results suggest a developmental divide between early and late childhood similar to that reported in some mainstream developmental research on analogical reasoning (e.g., Sternberg & Rifkin, 1979).

The conclusion that children in early childhood are unable to demonstrate analogy is disputed by other mainstream researchers (see e.g., Goswami, 1991). However, in more recent work, RFT researchers have provided evidence that suggests that apparently successful analogical responding as demonstrated by children younger than 5 years of age might be based on processes other than matching functionally same relations (Carpentier, Smeets, Barnes-Holmes & Stewart, 2004). This additional evidence supports the original conclusion from RFT research as well as that of the developmental work with whose conclusions it agreed.

More recent studies with adults have provided additional evidence that equivalence-equivalence or coordinate framing of relations provides a good model of analogical reasoning. Barnes-Holmes et al. (2005 a & b) examined event-related potentials associated with coordinate framing of both same and difference relations and reported patterns of neural activity paralleling those seen in studies that have examined the neural substrates of natural language analogy. Ruiz and Luciano (2011) extended the RFT model of analogy by training and testing "cross-domain" analogy as the relation of relations in separate relational networks. More importantly from the current perspective, performance on this model of analogy strongly correlated with that on a standard measure of analogical reasoning.

Perspective Relations

Relational frame theory approaches perspective-taking as arbitrarily applicable relational responding under the control of the (deictic) contextual cues I-YOU, HERE-THERE and NOW-THEN, or their functional equivalents. Published research examining this form of responding first emerged only in 2004 but there has been substantial additional work since then showing the correlation between deictic framing and perspective-taking as conceptualized in the mainstream literature as well as cognitive ability more generally.

McHugh, Barnes-Holmes and Barnes-Holmes (2004) investigated the development of perspective-taking as deictic framing in a range of age groups from early childhood to adulthood. Findings indicated that accuracy increased as a function of age, supporting the concept of the operant nature of this phenomenon. In addition, findings showed overlap with those of mainstream developmental as well as *Theory of Mind* (ToM) research. Regarding the former, they found that HERE-THERE relations tend to emerge before NOW-THEN relations, which coheres with evidence that young children master spatial before temporal relations. With respect to Theory of Mind, the literature in this domain argues that performance on simple ToM tasks should improve between four and five years old (Baron-Cohen, Tager-Flusberg, & Cohen, 2000). The findings from McHugh et al. in which the performances of children in their middle childhood more closely resembled those of older participants than those of the youngest age group accorded with this claim.

There is also evidence from studies of individuals in clinical categories associated with deficits in perspective-taking ability that demonstrate the correlation between deictic framing performance and the latter. Rehfeldt, Dillen, Ziomek and Kowalchuk (2007) found that children with autistic spectrum disorder (ASD) showed deficiencies in deictic framing in comparison with typically developing children. They also showed that accuracy on NOW-THEN framing correlated with scores on the Daily Living Skills domain of the Vineland Adaptive Behavior Scales, a standardized instrument commonly used in the assessment of ASD. Villatte, Monestes, McHugh, Freixa i Baque & Loas (2010a) found poorer performance of participants with schizophrenia than age matched controls in responding in accordance with complex deictic framing tasks. They also found that performance on the latter was a strong predictor of accuracy on the so-called mental states attribution task, a mainstream measure of perspective-taking. Villatte, Monestes, McHugh, Freixa i Baque & Loas (2010b) found poorer performance of both clinical participants with schizophrenia and sub-clinical participants with social anhedonia on deictic framing tasks.

The results just described further support the contention that deictic relational framing is a core process underlying perspective-taking ability. One other recent study shows the correlation between deictic framing and cognitive performance more generally. Gore, Barnes-Holmes and Murphy (2010) exposed 24 adults with varying levels of intellectual disability to standard measures of language and IQ, as well as to an adaptation of the McHugh et al. (2004) deictic framing protocol and found that perspective-taking correlated with verbal ability, full-scale IQ and performance IQ.

Training Relational Framing Repertoires

The research just reviewed shows how specific forms of framing are linked with specific cognitive abilities as well as how framing in general correlates with linguistic and cognitive ability. This supports the RFT contention that relational framing is the core repertoire underlying intellectual performance and suggests that by targeting this repertoire for training, we can improve performance in educational and other domains in which intellectual performance is important. Furthermore, there is now direct evidence of this. In what follows we will review RFT research on multiple exemplar training of a variety of relational frames.

Coordination

The earliest published work to report the explicit use of MET to establish derived relations was by Barnes-Holmes, Barnes-Holmes, Roche and Smeets (2001a). This study showed that MET could facilitate a transfer of function via symmetrical relations in 4-5 year old children. Across Experiments 1-3, twelve participants were trained to name a range of actions and objects and were also trained in action-object conditional discriminations (e.g., when experimenter waves, choose toy car; when experimenter claps, choose doll) and tested for derived object-action symmetry (e.g., given toy car, wave; given doll, clap), using a multiple baseline design to introduce MET in symmetrical responding for participants who failed. The use of the multiple baseline design provided evidence that MET and not name training was the important variable in facilitating symmetry. In Experiment 4, an additional four participants were trained and tested in the reverse direction (i.e., object-action conditional discrimination training before testing for action-object symmetry). A subsequent study (Barnes-Holmes, Barnes-Holmes, Roche & Smeets, 2001b) showed that MET was equally facilitative of derived symmetry in the absence of name training.

A number of further studies have added to the evidence of the efficacy of MET for establishing derived coordinate relations. Murphy, Barnes-Holmes and Barnes-Holmes (2005) used exemplar training to establish transfer of mand functions through equivalence relations in a child with ASD when he alone of six children failed to show the derived performance. Gomez, Lopez, Martin, Barnes-Holmes & Barnes-Holmes (2007) extended the work of Barnes-Holmes, Barnes-Holmes, Roche and Smeets (2001a & b), who had previously demonstrated the use of MET to facilitate mutually entailed coordinate framing in 4-year old children, by successfully employing it to facilitate combinatorial entailed coordinate framing in a similar age group. Luciano, Gomez-Becerra and Rodriguez-Valverde (2007) investigated the respective roles of MET and naming in the emergence over several months of immediate and delayed symmetrical responding and the emergence of equivalence in the context of two and three comparison formats in a very young infant (starting age: 15 months, 24 days; age by study end: 23 months, 25 days). The repertoires targeted by exemplar training emerged in the absence of naming, which only appeared by the end of the study. Murphy & Barnes-Holmes

(2010) used MET to establish derived manding in an adolescent with ASD, thus providing another empirical demonstration of the effect first shown by Murphy, Barnes-Holmes and Barnes-Holmes (2005). Finally, most recently, Cassidy, Roche and Hayes (2011) used MET to train typically developing 8-12 year old children in equivalence relations and showed not only an improvement in equivalence responding itself but also in full and subscale IQ scores.

Comparison & Opposition

There have been a number of studies that have used MET for the purposes of training non-coordinate relations. The two relations that have received most of the focus are comparison and opposition. This is unsurprising since these relations are of fundamental importance and likely provide basic templates for aspects of more complex relations. For example, comparison arguably provides a basic template that lends itself to features of spatial, temporal and hierarchical relations, while opposition arguably provides a similar template with respect to deictic relations. Hence, training these relations not only teaches the relations themselves but also lays the ground for more advanced relational learning. In addition, these relations are more complex than coordination and hence can provide a more rigorous training regimen in terms of improving cognitive performance.

The first study to train non-coordinate framing was Barnes-Holmes, Barnes-Holmes, Smeets, Strand and Friman (2004), who focused on comparative relations. This study also aspired to be the first to generate repertoires of relational responding when found to be absent in young children, as opposed to simply facilitating them as had likely happened in Barnes-Holmes, Barnes-Holmes, Roche and Smeets (2001a & b). In Barnes-Holmes, Barnes-Holmes, Smeets et al. (2004), three children between 4 and 6 years were exposed to a task involving two or three identically-sized paper "coins". On each trial, the experimenter would compare the coins in terms of value and then ask the child to pick the one that would "buy as many sweets as possible." All three children failed baseline tests for specific patterns of comparative framing. Hence, MET was employed to train the appropriate, increasingly complex patterns required. Generalization tests showed that the taught patterns generalized to novel stimuli and a novel experimenter. Furthermore, evidence of the operant nature of the repertoire being trained was provided on the basis of a noted absence of improvement under conditions of non-contingent reinforcement and by demonstrating contingency reversals for all three children.

The second study to train non-coordinate framing focused on opposition relations. This study, by Barnes-Holmes, Barnes-Holmes and Smeets (2004), was a companion study to the one just reviewed, in that it employed the same three children and used a similar task to test and train opposition relations. Children were presented with various numbers of paper "coins" whose value relative to other coins was specified in each trial via relations of opposition and were asked which coin or coins they would choose in order to "buy as many sweets as possible." All three children failed baseline tests and were subsequently trained to respond appropriately through MET. Generalization and evidence of the operant nature of the repertoire were seen in this case as for comparison.

Berens and Hayes (2007) replicated and extended Barnes-Holmes, Barnes-Holmes, Smeets, Strand and Friman (2004) on MET of comparative relations by using a combined multiple baseline and multiple probe design to conduct a more thorough analysis of the effects of MET in the context of comparative relations. A multiple baseline (across responses) design allowed the assessment of the effects of training of an array of increasingly complex aspects of the comparative relational operant (i.e., mutually entailed more, mutually entailed less, combinatorial entailed more, combinatorial entailed less, mixed non-linear more and less) on each of these aspects themselves as well as on the complete comparative frame. The inclusion of the final aspect of the comparative operant, mixed non-linear comparative relations, facilitated examination of a more advanced comparative repertoire. Finally, the use of extended baselines in the context of a multiple baseline (across participants) design allowed more clear cut evidence that the procedure was establishing the comparative frame rather than simply establishing a novel context for an already established operant.

Two recent studies that used MET to train comparative relations in individuals with developmental delay have added to evidence of the efficacy of MET for training comparative relations and relational frames more generally and provided further support for the use of RFT-based interventions for individuals with developmental delay. Gorham,

Barnes-Holmes, Barnes-Holmes, and Berens (2009) used MET to train both typically developing children and children with autism to respond correctly on relatively advanced comparative relational tasks involving 4 "coin" stimuli, while Murphy and Barnes-Holmes (2010) used MET to train transformation of manding functions through derived comparative relations in one typically developing child and one child with autism.

The last and most recent example of a study that falls under the current heading involved the training of both the non-coordinate relations under discussion (i.e., comparison and opposition) and is perhaps one of the most convincing studies to have emerged thus far in respect of the potential of RFT to improve intellectual ability and further educational aims. Cassidy et al. (2011) were already reported in the sub-section on coordinate relations as having used MET to successfully train 8-12 year olds in equivalence relations. In fact, they trained separate groups of educationally typical and educationally sub-typical children not just in equivalence relations but also in other relations including comparison and opposition. In the case of all three types of relations, multiple exemplar training successfully improved relational framing. Furthermore, as previously indicated, relational training produced improvements in full and subscale IQ. While, as reported, equivalence training proved to be relatively beneficial in this respect, training in multiple non-coordinate relations seemed to produce much greater improvements even than equivalence training. In Experiment 1, which focused on educationally typical children, for example, the effect size for improvement in full-scale IQ after equivalence training was 0.98, while the effect size for improvement in IQ after multiple relational training was 5.13. In Experiment 2, which involved training 8 children with educational difficulties in multiple relational training alone, full-scale IQ rose by at least one standard deviation for seven of the eight children, and this change was also significant at the group level. Furthermore, follow-up IQ testing conducted almost four years later showed that in all cases IQ rises were maintained well across this very large follow-up period (Roche, Cassidy & Stewart, in press). Overall, the findings from this study suggest the efficacy of MET of relational framing for the skill of relational framing itself as well as for intellectual performance, and point to the importance of non-coordinate framing in particular in this respect.

Perspective Relations

A recent study by Weil, Hayes, and Capurro (2011) used MET to enhance deictic framing in typically developing preschool children. Children were trained in multiple exemplars of I/You, Here/There, and Now/Then relations, and training progressed from simplest to most complex relations (i.e., from simple, to reversed, and finally to double-reversed relations). Training resulted in increases in accurate performance and generalization to untrained stimuli across participants. General increases were also seen post-training on cognitive tests of perspective-taking. This study was the first to demonstrate that MET can improve derived relational responding relevant to perspective-taking. An important next step will be investigation of MET as a method for training deictic relations in populations characterized by perspective–taking deficits, such as those with autism.

Sequential Relations

Performance on sequential relational response tasks has been shown to correlate with fluid intelligence (i.e., the ability to reason and solve new problems independently of previously acquired knowledge; Jäeggi, Studer-Luethi, Buschkuehl, Su, Jonides & Perrig, 2010). A recent study by Baltruschat, Hasselhorn, Tarbox, Dixon, Najdowski, Mullins, and Gould (2012) used MET to improve performance of children with autism on a sequential relational task known as the "digit span backward" task. In this task, the experimenter vocally presented a random sequence of letters to the child and the latter was then asked to repeat the letters back to the experimenter in reverse order. MET was used to improve accuracy on the task and improvements were seen in all children. Furthermore, increases in accuracy maintained after reinforcement was discontinued and generalized to untrained stimuli (letters and numbers that were not included in training).

Conditional Relations

In a recent study by Tarbox, Zuckerman, Bishop, Olive, and O'Hora (2011), MET was used to teach six children with autism to respond to rules that described conditional relations between antecedents and behaviors, a foundational component of rule-governed behavior. Experimenters presented rules in "If/then" format, for example, "If this is a carrot, then clap your hands." Trials randomly alternated between presenting stimuli specified in the rule (e.g., a carrot) versus stimuli that were not specified in the rule. Correct responding consisted of emitting the behavior specified in the rule when the specified antecedent stimulus was present and inhibiting the behavior when the specified antecedent was not present (e.g., not clapping if an apple was presented instead of a carrot). In baseline, none of the participants were able to consistently respond correctly. Prompting, reinforcement, and prompt fading were used to train multiple exemplars of particular antecedent-behavior combinations. After MET, all participants demonstrated generalization to numerous untrained rules (i.e., rules that contained antecedents and behaviors that were never present during training).

Other Relevant Work

Mathematical Relations

Chris Ninness and colleagues have published a series of articles reporting the successful use of computer-interactive protocols to teach advanced mathematical skills via derived relations (Ninness, Rumph, McCuller, Vasquez, Harrison, Ford, Capt, Ninness & Bradfield, 2005; Ninness, Rumph, McCuller, Harrison, Ford & Ninness, 2005; Ninness, Barnes-Holmes, Rumph, McCuller, Ford, Payne, Ninness, Smith & Ward & Elliott, 2006; Ninness, Dixon, Barnes-Holmes, Rehfeldt, Rumph, McCuller, Holland, Smith, Ninness, & McGinty, 2009). These studies involved using matching-to-sample strategies to (i) teach formula-to-graph relations for mathematical transformations (of algebraic and trigonometric formulae) about the coordinate axes (Ninness, Rumph, McCuller, Vasquez et al., 2005); (ii) teach formula-to-factored formula and factored formula-to-graph relations for vertical and horizontal shifts on the coordinate axes (Ninness, Rumph, McCuller, Harrison et al., 2005); (iii) analyze and guide transformation of stimulus functions of formulae (Ninness et al., 2006); and (iv) incorporate same and opposite (reciprocal) relational control into training of trigonometric classes involving transformation of amplitude and frequency functions (Ninness et al., 2009).

Metaphorical Relations

Metaphor is a subtype of analogy to the extent that it involves relating relational networks. However, metaphors are often more subtle and complex than analogies. In addition, they are relatively frequent in naturalistic language contexts and thus understanding and using them is an important skill. Another recent study on MET taught children with autism to solve novel metaphors (Persicke, Tarbox, Ranick, & St Clair, 2012). Children with autism were taught to solve novel metaphors (e.g., "This apple is candy") by talking themselves through the following steps: 1) stating the properties of the apple (e.g., it's fruit, it grows on trees, it's sweet), 2) stating the properties of the metaphor (e.g., "it's food, it's sweet, and it rots your teeth"), and 3) identifying the property shared between the two stimuli, thereby identifying the meaning of the metaphor (in this case it means the apple tastes sweet). From an RFT perspective, the children were being taught to relate each stimulus in the metaphor to its properties; instances of hierarchical relating. In addition, comparing the inferred properties of the two stimuli in the metaphor consisted of relational responding in terms of both differentiation (for properties that were not the same) and coordination (when the particular shared property was identified). All children in the study learned to solve novel metaphors and accurate responding generalized to metaphors that had never been trained, thereby demonstrating a flexible, generally applicable repertoire.

Overview and Conclusion

The research reviewed in this chapter provides significant evidence that relational framing (i) correlates with intellectual skills and abilities required to succeed in the educational arena; (ii) can be treated as operant behavior, and, most critically for educational purposes, strengthened via multiple exemplar training; and (iii) has substantial effects on measured intellectual ability when it is trained. The latter work in particular promises great progress in the applied educational realm if RFT procedures were to be applied systematically therein.

This work represents a considerable advance on that reported in the first edited book on RFT. Recent research showing that relational framing can be trained and that such training can have positive effects on cognitive ability is particularly exciting. For example, the study by Cassidy et al. (2011) showing the effect of relational framing on IQ is one of the most promising studies to emerge from the RFT stable thus far in respect of the impact of MET on educationally important outcomes. Nevertheless, much work remains to be done. As regards the Cassidy et al. study, for instance, the effects shown were obtained with a small sample and with less rigorous standards of control than might be desirable. Hence, larger scale and more tightly controlled versions of this kind of research are needed.

Future RFT research focused on intellectual development and education should also consider the use of additional dependent measures. Many of the studies reviewed here have used IQ and standardized language instruments as correlational measures. IQ in particular has been used as a surrogate for other educational outcomes. There is good reason for this since IQ as a measure of intellectual functioning correlates highly with educational outcomes (Deary, Strand, Smith & Fernandes, 2007). However, research involving other variables that are of more obviously direct educational relevance, including for example academic or career performance, is needed. This will be particularly important with respect to RFT work on the effect of MET for relational framing. Information on these other outcome variables may necessitate longitudinal research. Nevertheless such work is crucial in providing a valid assessment of impact on educational success.

Apart from providing more thorough investigation of the effects of relational framing in general on educational outcomes, there is also more detailed information needed on framing itself. Though we have acquired substantial evidence of the importance of framing in general to language and cognition, research into framing at a more detailed level is still at a relatively early stage. For example, there remains as yet little or no investigation of some unquestionably important forms of framing (e.g., spatial, temporal, logical or hierarchical) and as yet limited investigation of some others (e.g., conditional, sequential) or how relational frames in general interact with and support each other in the course of development.

Building up a detailed picture of the typical development of each of the major frames and how they interact with each other will be of major educational benefit. In addition, research into relational framing up to this point has focused solely on latency and stimulus control as dependent variables. However, intelligent behavior is characterized by a number of different features including "range, speed, flexibility and subtlety of contextual control over relational responding" (Barnes-Holmes, Barnes-Holmes, Roche, Healy, Lyddy, Cullinan & Hayes, 2001, p. 161). Hence, while latency and stimulus control are important, additional emphasis on alternative features of intelligent performance including especially flexibility and subtlety of contextual control is needed.

Finally, there is also much more work needed to adapt RFT training procedures for applied educational domains. As the current chapter suggests, important inroads are beginning to be made in this respect, but the research to date merely scratches the surface in terms of assessment and training of relational framing. Given the tremendous potential of RFT to facilitate intellectual development and boost achievement of educational outcomes, it behooves those in the RFT community to translate this approach into the applied educational arena as rapidly and effectively as possible.

References

Baltruschat, L., Hasselhorn, M., Tarbox, J., Dixon, D. R., Najdowski, A. C., Mullins, R. D., & Gould, E. R. (2012). The effects of multiple exemplar training on a working memory task involving sequential responding in children with autism. *The Psychological Record, 62*, 549.

Barnes, D., Hegarty, N., & Smeets, P. M. (1997). Relating equivalence relations to equivalence relations: A relational framing model of complex human functioning. *The Analysis of Verbal Behavior, 14*, 1-27.

Barnes, D., McCullagh, P., & Keenan, M. (1990). Equivalence class formation in non-hearing impaired children and hearing impaired children. *Analysis of Verbal Behavior, 8*, 1-11.

Barnes-Holmes, Y., Barnes-Holmes, D., & Cullinan, V. (2001). Education. In S. C. Hayes, D. Barnes-Holmes & B. Roche (Eds.), *Relational frame theory: A post-Skinnerian approach to language and cognition.* New York: Plenum Press.

Barnes-Holmes, Y., Barnes-Holmes, D., Roche, B., Healy, O., Lyddy, F., Cullinan, V., & Hayes, S.C. (2001). Psychological Development. In S.C. Hayes, D.

Barnes-Holmes, & B. Roche (Eds.) *Relational Frame Theory: A Post-Skinnerian Approach to Language and Cognition* (p. 161). New York: Plenum Press.

Barnes-Holmes, Y., Barnes-Holmes, D., Roche, B., & Smeets, P. M. (2001a). Exemplar training and a derived transformation of functions in accordance with symmetry. *The Psychological Record, 51,* 287-308.

Barnes-Holmes, Y., Barnes-Holmes, D., Roche, B., & Smeets, P. M. (2001b). Exemplar training and a derived transformation of functions in accordance with symmetry: II. *The Psychological Record, 51,* 589-604.

Barnes-Holmes, Y., Barnes-Holmes, D., & Smeets, P. M. (2004). Establishing relational responding in accordance with opposite as generalized operant behavior in young children. *International Journal of Psychology and Psychological Therapy, 4,* 559-586.

Barnes-Holmes, Y., Barnes-Holmes, D., Smeets, P. M., Strand, P., & Friman, P. (2004). Establishing relational responding in accordance with more-than and less-than as generalized operant behavior in young children. *International Journal of Psychology and Psychological Therapy, 4,* 531-558.

Barnes-Holmes, D., Staunton, C., Whelan, R., Barnes-Holmes, Y., Commins, S., Walsh, D., Stewart, I., Smeets, P. M., & Dymond, S. (2005a). Derived stimulus relations, semantic priming, and event-related potentials: Testing a behavioral theory of semantic networks. *Journal of the Experimental Analysis of Behavior. 84,* 417-434.

Barnes-Holmes, D., Regan, D., Barnes-Holmes, Y., Commins, S., Walsh, D., Stewart, I., Smeets, P. M., Whelan, R., & Dymond, S. (2005b). Relating derived relations as a model of analogical reasoning: Reaction times and event related potentials. *Journal of the Experimental Analysis of Behavior. 84,* 435-452.

Baron-Cohen, S., Tager-Flusberg, H., & Cohen, D. (Eds.). (2000). *Understanding other minds: perspectives from developmental cognitive neuroscience.* Oxford University Press.

Berens, N. M., & Hayes, S. C. (2007). Arbitrarily applicable comparative relations: Experimental evidence for relational operants. *Journal of Applied Behavior Analysis, 40,* 45-71.

Carpentier, F., Smeets, P. M., & Barnes-Holmes, D. (2002). Matching functionally-same relations: Implications for equivalence-equivalence as a model for analogical reasoning. *The Psychological Record, 52,* 351-370.

Carpentier, F., Smeets, P. M., & Barnes-Holmes, D. (2003). Equivalence-equivalence as a model of analogy: Further analyses. *The Psychological Record, 53,* 349-372.

Carpentier, F., Smeets, P. M., Barnes-Holmes, D., & Stewart, I. (2004). Matching derived functionally-same relations: Equivalence-equivalence and classical analogies. *The Psychological Record, 54,* 255-273.

Cassidy, S., Roche, B., & Hayes, S. C. (2011). A relational frame training intervention to raise intelligence quotients: A pilot study. *The Psychological Record, 61,* 173-198.

Deary, I. J., Strand, S., Smith, P. & Fernandes, C. (2007). Intelligence and educational achievement, *Intelligence, 35*, 13-21.

Dickins, D. W., Singh, K. D., Roberts, N., Burns, P., Downes, J. J., Jimmieson, P., & Bentall, R. P. (2001). An fMRI study of stimulus equivalence. *Neuroreport, 12(2)*, 405-411.

Dugdale, N., & Lowe C. F. (2000). Testing for symmetry in the conditional discriminations of language-trained chimpanzees. *Journal of the Experimental Analysis of Behavior, 73*, 5-22.

Gomez, S., Lopez, F., Martin, C. B., Barnes-Holmes, Y., & Barnes-Holmes, D. (2007). Exemplar training and derived transformation of functions in accordance with symmetry and equivalence. *The Psychological Record, 57*, 273-294.

Gore, N. J., Barnes-Holmes, Y., & Murphy, G. (2010). The relationship between intellectual functioning and relational perspective-taking. *International Journal of Psychology and Psychological Therapy, 10(1)*, 1-17.

Gorham, M., Barnes-Holmes, Y., Barnes-Holmes, D. & Berens, N. (2009). Derived comparative and transitive relations in young children with and without autism. *The Psychological Record, 59*, 221-246.

Goswami, U. (1991). Analogical reasoning: What develops? A review of research and theory. *Child Development, 62*, 1-22.

Greer, D. (1991). The teacher as strategic scientist: A solution to our educational crisis? *Behavior and Social Issues, 1(2)*, 25-41.

Hayes, S. C., & Bissett, R. (1998). Derived stimulus relations produce mediated and episodic priming. *The Psychological Record, 48*, 617-630.

Hayes, S.C., Fox, E., Gifford, E., Wilson, K., Barnes-Holmes, D., & Healy, O. (2001). Derived relational responding as learned behavior. In Hayes, S.C., D. Barnes-Holmes & B.Roche (Eds.) *Relational Frame Theory: A Post Skinnerian Account of Human Language and Cognition* (p.36). New York: Plenum Press.

Jäeggi, S. M., Studer-Luethi, B., Buschkuehl, M., Su, Y., Jonides, J. & Perrig, W. J. (2010). The relationship between n-back performance and matrix reasoning implications for training and transfer. *Intelligence, 38*, 625-635.

Leader, G., & Barnes-Holmes, D. (2001). Establishing fraction-decimal equivalence using a respondent-type training procedure. *The Psychological Record, 51*, 151-166.

Luciano, C., Gómez-Becerra, I., & Rodríguez-Valverde, M. (2007). The role of multiple-exemplar training and naming in establishing derived equivalence in an infant. *Journal of the Experimental Analysis of Behavior, 87*, 349-365.

McHugh, L., Barnes-Holmes, Y., & Barnes-Holmes, D. (2004). A relational frame account of the development of complex cognitive phenomena: Perspective-taking, false belief understanding, and deception. *International Journal of Psychology and Psychological Therapy, 4*, 303-323.

Moran, L., Stewart, I., McElwee, J. & Ming, S. (2010). The training and assessment of relational precursors and abilities (TARPA): A preliminary analysis. *Journal of Autism and Developmental Disorders 40(9)*, 1149-1153.

Munnelly, A., Dymond, S., & Hinton, E. C. (2010). Relational reasoning with derived comparative relations: A novel model of transitive inference. *Behavioral Processes, 85*, 8-17.

Murphy, C., & Barnes-Holmes, D. (2009a). Establishing derived manding for specific amounts with three children: An attempt at synthesizing Skinner's Verbal Behavior with Relational Frame Theory. *The Psychological Record, 59*, 75-92.

Murphy, C., & Barnes-Holmes, D. (2009b). Derived more-less relational mands in children diagnosed with autism. *Journal of Applied Behavior Analysis, 42*, 253-268.

Murphy, C., & Barnes-Holmes, D. (2010). Establishing complex derived manding with children with and without a diagnosis of autism. *The Psychological Record, 60*, 489-504.

Murphy, C., Barnes-Holmes, D., & Barnes-Holmes, Y. (2005). Derived manding in children with autism: Synthesizing Skinner's Verbal Behavior with relational frame theory. *Journal of Applied Behavior Analysis, 38*, 445-462.

Ninness, C., Barnes-Holmes, D., Rumph, R., McCuller, G., Ford, A., Payne, R., et al. (2006). Transformation of mathematical and stimulus functions. *Journal of Applied Behavior Analysis, 39*, 299–321.

Ninness, C., Dixon, M., Barnes-Holmes, D., Rehfeldt, R.A., Rumph, R., McCuller, G., Holland, J., Smith, R., Ninness, S. K., & McGinty, J. (2009). Constructing and deriving reciprocal trigonometric relations: A functional analytic approach. *Journal of Applied Behavior Analysis, 42(2)*, 191-208.

Ninness, C., Rumph, R., McCuller, G., Harrison, C.,Ford, A. M., & Ninness, S. (2005). A functional analytic approach to computer-interactive mathematics. *Journal of Applied Behavior Analysis, 38*, 1–22.

Ninness, C., Rumph, R., McCuller, G., Vasquez, E., Harrison, C., Ford, A. M., et al. (2005). A relational frame and artificial neural network approach to computer-interactive mathematics. *The Psychological Record, 51*, 561–570.

O'Connor, J., Rafferty, A., Barnes-Holmes, D., & Barnes-Holmes, Y. (2009). The role of verbal behavior, stimulus nameability, and familiarity on the equivalence performances of autistic and normally-developing children. *The Psychological Record, 59*, 53-74.

O'Hora, D., Pelaez, D., & Barnes-Holmes, D. (2005). Derived relational responding and performance on the verbal subtests of the WAIS III. *The Psychological Record, 55*, 155-175.

O'Hora, D., Pelaez, M., Barnes-Holmes, D., Rae, G., Robinson, K., & Chaudhary, T. (2008). Temporal relations and intelligence: Correlating relational performance with performance on the WAIS-III. *The Psychological Record, 58*, 569-584.

O'Toole, C., & Barnes-Holmes, D. (2009). Three chronometric indices of relational responding as predictors of performance on a brief intelligence test: The importance of relational flexibility. *The Psychological Record, 59*, 119-132.

Persicke, A., Tarbox, J., Ranick, J., & St Clair, M. (2012). Establishing metaphorical reasoning in children with autism. *Research in Autism Spectrum Disorders, 6*, 913-920.

Rehfeldt, R. A., Dillen, J. E., Ziomek, M. M., & Kowalchuk, R. E. (2007). Assessing relational learning deficits in perspective-taking in children with high functioning autism spectrum disorder. *The Psychological Record, 57*, 23-47.

Rehfeldt, R. A., & Root, S. L. (2005). Establishing derived requesting skills in adults with severe developmental disabilities. *Journal of Applied Behavior Analysis, 38*, 101-105.

Reilly, T., Whelan, R., & Barnes-Holmes, D. (2005). The effects of training structure on the latency of responses to a five-term linear chain. *The Psychological Record, 55*, 233-249.

Roche, B., Cassidy, S. & Stewart, I. (in press). Nurturing genius: Realizing a foundational aim of Psychology. In T. Kashdan & J. Ciarrochi (Eds.), *Cultivating well-being: Treatment innovations in Positive Psychology, Acceptance and Commitment Therapy, and beyond.* Oakland, CA: New Harbinger.

Ruiz, F. J. & Luciano, C. (2011). Cross-domain analogies as relating derived relations among two separate relational networks. *Journal of the Experimental Analysis of Behavior 95(3)*, 369-385.

Smith, B. L., Smith, T. D., Taylor, L. & Hobby, M. (2005). Relationship between intelligence and vocabulary. *Perception and Motor Skills, 100(1)*, 101-108.

Sternberg, R. J., & Rifkin, B. (1979). The development of analogical reasoning processes. *Journal of Experimental Child Psychology, 27*, 195–232.

Tarbox, J., Zuckerman, C. K., Bishop, M. R., Olive, M. L., & O'Hora, D. P. (2011). Rule-governed behavior: Teaching a preliminary repertoire of rule-following to children with autism. *The Analysis of Verbal Behavior, 27*, 125-139.

Villatte, M., Monestès, J. L., McHugh, L., Freixa i Baqué, E., & Loas, G. (2010a). Assessing perspective taking in schizophrenia using Relational Frame Theory. *The Psychological Record, 60*, 413-424.

Villatte, M., Monestès, J. L., McHugh, L., Freixa i Baqué, E., & Loas, G. (2010b). Adopting the perspective of another in belief attribution: Contribution of Relational Frame Theory to the understanding of impairments in schizophrenia. *Journal of Behavior Therapy and Experimental Psychiatry, 41*, 125-134.

Weil, T. M., Hayes, S. C., & Capurro, P. (2011). Establishing a deictic relational repertoire in young children. *The Psychological Record, 61*, 371-390.

CHAPTER NINE

Relational Frame Theory and Experimental Psychopathology

Simon Dymond

Swansea University

Bryan Roche

National University of Ireland, Maynooth

Marc Bennett

University of Leuven

I t is now over a decade since the publication of the "purple RFT book" (Hayes, Barnes-Holmes, & Roche, 2001), and since then empirical research into RFT processes and applications has grown dramatically (Dymond, May, Munnelly, & Hoon, 2010). As the contents of the present volume testify, research on RFT has advanced the functional contextualistic understanding of language and cognition in numerous ways. An important research domain for the further extension and analysis of RFT is experimental psychopathology and, in particular, understanding the role of verbal relational processes in the myriad complex behaviors involved in human suffering.

Despite the acknowledged link between RFT and Acceptance and Commitment Therapy (ACT; Hayes, Luoma, Bond, Masuda, & Lillis, 2006), basic empirical research on the putative behavioral processes involved in the acquisition and maintenance of clinically relevant behavior has perhaps not progressed at a pace one might have expected, at least not compared to the dissemination of ACT methods and evaluations (Blackledge & Drake, this volume). Rather than speculate on the possible reasons for the slow growth of RFT research in experimental psychopathology we will instead review the existing research on RFT analyses of clinically relevant behavior and outline several issues and topics worthy of further investigation. The challenge, then, is to show that RFT *can* be applied to the analysis of clinically significant behavior both in the applied setting and in the laboratory, to outline empirical support for an RFT approach accumulated in the years since the "RFT book", and to emphasize some points of overlap with other approaches to clinically relevant behavior.

The RFT View of Psychopathology

RFT contends that psychopathology, and most forms of human suffering, are the products of arbitrarily applicable relational framing (Dymond & Roche, 2009). The seemingly unique human ability to be haunted by a painful remembered past or to worry about an anticipated future, which is the flip side of the relational self-knowledge coin (Dymond & Barnes, 1997), is "a primary source of psychopathology (as well as a process exacerbating the impact of other sources of psychopathology)" [and interacts] "with direct contingencies to produce an inability to persist or change behavior in the service of long term valued ends" (Hayes et al., 2006, p. 6). Relational framing, then, is a core process in acquiring and maintaining psychopathology that underpins the embedded repertoires of behavioral inflexibility illustrated by the ACT/RFT model (Blackledge & Drake, this volume).

While overlap with other views of psychopathology may be noted (Bentall, 2006; Helzer, Kraemer & Krueger, 2006; van Os, Linscott, Myin-Germeys, Delespaul, & Krabbendam, 2009), such accounts are hamstrung by the absence of a functional account of verbal processes. On the other hand, RFT has a clear, technical definition of verbal events as those that have their functions based at least in part on participation in relational frames (Dymond & Rehfeldt, 2000; Dymond & Roche, 2009). As we will now see, this RFT approach to psychopathology has already developed a respectable evidence base.

We will now outline advances made in RFT research on fear conditioning, extinction, avoidance, specific phobia, and thought suppression. The research reviewed in this chapter does not include component studies of the efficacy of ACT, of which there are many excellent reviews available (e.g., Blackledge & Drake, this volume; Hayes et al., 2006). These studies examine the effectiveness of ACT or ACT components on a variety of clinical outcome measures with clients in therapy. However, while the interventions assessed in ACT component studies are informed by RFT, they do not assess the occurrence of relational framing *in vivo* or examine relational processes directly. In other words, ACT component studies can support an RFT *interpretation* of clinical phenomena, but they cannot easily show that the relevant relational processes (e.g., transformation of functions) are actually occurring during the therapeutic process.

Another variety of study comes closer to the types of rigorous analysis of psychopathology processes of interest in the current chapter. These studies use non-clinical population samples and typically correlate the probabilities of some clinically analogous behavior, such as avoidance of spiders (e.g., Cochrane, Barnes-Holmes, & Barnes-Holmes, 2008), responses to a challenging perceptual motor task (Zettle, Petersen, Hocker, & Provines, 2007), or frequency of internet pornography use (e.g., Wetterneck, Burgess, Short, Smith, & Cervantes, 2012) with measures of such constructs as experiential avoidance (e.g., the *Acceptance and Action Questionnaire*; Bond et al., 2011). While these studies come closer to experimental control over the behaviors of interest, they nevertheless remain correlational and interpretative in their investigations of relational processes. Closer still to the "gold standard" of experimental control of interest here are studies that employ non-clinical samples and even computer delivered therapeutic interventions to examine, for instance, the effects of verbal instructions on analogs of problem behavior, such as avoidance of an aversive event such as pain (e.g., Gutiérrez, Luciano, Rodríguez, & Fink, 2004; Keogh, Bond, Hanmer, & Tilston, 2005; McMullen, Barnes-Holmes, Barnes-Holmes, Stewart, & Cochrane, 2007; Roche, Forsyth & Maher, 2007; Wagener & Zettle, 2011; see also Masuda, Hayes, Sackett & Twohig, 2004). These studies achieve impressive levels of control over the manipulation of the intervention protocols and the measurement of study outcomes. However, they still suffer from the limitation that they do not directly examine relational processes at work during the emergence of the analogous problem behavior (e.g., avoidance or an ineffective coping strategy) or during the therapeutic intervention.

What we are most interested in here are studies that complement those of the foregoing kind, by obtaining full experimental control over the relational processes themselves, during the emergence and analog treatment of clinically relevant behavior. Such studies employ relational frame training procedures, in order to establish derived relational networks, and transform behavioral/emotional response functions in accordance with those networks to both establish and diminish behaviors analogous of psychopathology. While the body of published studies of this latter kind is growing, they are still in relatively short supply and focus almost exclusively on the generation and analog treatment of fear and avoidance.

Relational Frame Theory and Experimental Analysis of Fear and Avoidance

Derived Fear and Avoidance

Fear conditioning is a central topic in experimental psychopathology, with a vast literature on basic and clinical features of the acquisition of fear-eliciting functions by previously neutral stimuli (Beckers, Krypotos, Boddez, Effting, & Kindt, 2013). Recently, a great deal of attention has been paid to the generalization of fear-provoking functions in both clinical (e.g., Lissek et al., 2010) and nonclinical participants (e.g., Dunsmoor, Mitroff, & LaBar, 2009; Vervliet, Kindt, Vansteenwegen, & Hermans, 2010). Researchers in the associative learning tradition accept that while clinical and anecdotal reports of anxiety often describe a prior experience of fear conditioning that precipitates avoidance, direct contact with an aversive event is often not necessary for fear conditioning and avoidance to occur (Askew & Field, 2008; Muris & Field, 2010; Olsson & Phelps, 2004, 2007). For instance, it has been found that stimulus generalization may occur whereby stimuli physically similar to a laboratory conditioned fear stimulus (CS) occasion responding along a specified dimension. That is, fear conditioning has been shown to generalize along perceptual (Dunsmoor et al., 2009; Vervliet et al., 2010) and conceptual/ semantic continua (Dunsmoor, Martin, & LaBar, 2012; Dunsmoor, White, & LaBar, 2011). In one study, Lommen, Engelhard, and van den Hout (2010) showed generalized avoidance of degraded colored circles along a physical continuum between CS+ and CS-. While this research is novel and methodologically rigorous, the dominant paradigm is associative and interpretations of study outcomes are usually mediational (e.g., Lovibond, 2006). In effect, much remains to be ascertained regarding the functional processes involved in the "symbolic" generalization of fear and avoidance along arbitrary continua.

Behavioral research into the derived transformation of fear and avoidance has shown that Pavlovian fear eliciting functions may both transfer (Dougher, Augustson, Markham, Greenway, & Wulfert, 1994; Rodriguez Valverde, Luciano, & Barnes-Holmes, 2009) and transform (Dougher, Hamilton, Fink, & Harrington, 2007; Roche & Barnes, 1997; Roche, Barnes-Holmes, Smeets, Barnes-Holmes, & McGeady, 2000) in accordance with relational frames, although such effects warrant replication and extension (see also, Stewart & McAlwee, 2009). This phenomenon has been used as a behavioral interpretive paradigm within which generalization of avoidance and fear along arbitrary stimulus continua can be understood without appeal to mentalistic processes (Augustson & Dougher, 1997; Dymond, Roche, Forsyth, Whelan, & Rhoden, 2007, 2008; Dymond, Schlund, Roche, Whelan, Richards, & Davies, 2011; Roche, Kanter, Brown, Dymond, & Fogarty, 2008; Smyth, Barnes-Holmes, & Forsyth, 2006). In one early study of derived avoidance, Augustson and Dougher (1997) first trained and tested participants for the formation of equivalence relations (AV1-AV2-AV3-AV4 and N1-N2-N3-N4). Next, in a differential conditioning procedure, one stimulus (AV2) was followed by shock and another (N2) was not. During the subsequent avoidance-learning phase, AV2 was followed by shock unless a fixed-ratio 20-response requirement was met, in which case AV2 was removed from the screen and the scheduled shock omitted. The transfer of this threat-avoidance responding was then tested with presentations of stimuli that had not been present during the avoidance-learning phase. Findings showed that all participants emitted the derived threat-avoidance response to AV3 and AV4 (indirectly related to AV2) and not to N3 and N4 (indirectly related to N2).

Further studies have replicated and extended this basic effect. Dymond et al. (2007, 2008) demonstrated transformation of avoidance response function in accordance with same and opposite relational frames. The authors first established contextual functions of same and opposite for two disparate stimuli using nonarbitrary relational training and testing. Next, the following arbitrary stimulus relations were trained: Same-A1-B1, Same-A1-C1, Opposite-A1-B2, Opposite-A1-C2, which lead to the following derived relations: Same B1-C1, Same B2-C2, Opposite B1-C2 and Opposite B1-C1. Dymond et al. (2007, 2008) then exposed participants to a signaled avoidance task, during which responding in the presence of the stimulus B1 canceled a scheduled aversive image and sound. Another stimulus from the relational network, B2, was never followed by images or sounds. Once this avoidance response was acquired, participants were exposed to a probe phase in which C1 and C2 were presented in extinction. The majority of participants produced consistent avoidance responses in the presence of C1 but not C2 (i.e., C1 is the same as B1, whereas C2 is the opposite), thus demonstrating the transformation of avoidance response functions in accordance with complex relational networks.

A recent large-sample study by Dymond et al. (2011) showed transformation of avoidance functions in accordance with sameness when supplemental self-report measures were incorporated. Dymond et al. (2011) first trained and tested participants for the formation of stimulus equivalence relations (AV1-AV2-AV3 and N1-N2-N3). During the avoidance-learning phase, one stimulus (AV2) was followed by aversive images and sounds unless a response was made (pressing the space-bar once), and another (N2) was not. When the avoidance response was performed, AV2 was removed from the screen and the aversive stimuli omitted. The symbolic generalization of this threat-avoidance responding was then tested with presentations of stimuli that had not been present during the avoidance-learning phase. All participants readily made the threat-avoidance response to AV3 and not to N3. Additionally, measures of threat beliefs (subject's expectancies of aversive images following avoidance and non-avoidance) paralleled avoidance behavior patterns, thereby providing the first evidence of symbolic generalization of threat beliefs in accordance with a relational network.

All of the foregoing studies show that, at least in principle, it is possible for humans to develop avoidance patterns in the presence of novel stimuli in the absence of a direct history of (positive or negative) reinforcement for doing so. While we do not know the specific role played by the derived transformation of avoidance functions in the etiology of robust avoidance behavior observed in the clinical setting, it is possible that it plays at least an occasional part in the development of such problems.

More recently, researchers have begun to examine derived avoidance and simultaneous derived approach contingencies in an attempt to model more real-world instances of anxiety. That is, the anxiety condition comprises more than mere fear, or fear and avoidance combined. Rather, it has been suggested that for anxious clients approach and avoidance contingencies are likely working in combination with each other (Forsyth, Eifert, & Barrios, 2006; Hayes, 1976). That is, it may be the prevalence of competing approach (e.g., wanting to fly abroad for a holiday) and avoidance contingencies (e.g., fear of flying) that best characterizes the anxious state. Indeed, Hayes (1976) argued that without competing approach repertoires, an avoidance repertoire is arguably a functional rather than disordered response. In light of this, one recent study (Gannon, Roche, Kanter, Forsyth, & Linehan, 2011) juxtaposed competing approach and avoidance contingencies by presenting subjects with a stimulus that was equally related (i.e., in terms of nodal distance) to both a conditioned avoidance and a conditioned approach stimulus. Reaction times and response probabilities illustrated that subjects showed significantly delayed responding on juxtaposed approach-avoidance trials compared to approach only or avoidance only trials. Near perfect randomness in approach and avoidance probabilities on such trials were also observed across subjects.

Given the foregoing, it would appear that Relational Frame Theorists are more than well equipped to address the increasingly important questions regarding the emergence of fear and avoidance for stimuli not involved in prior conditioning. Indeed, the "discovery" and analysis of derived fear and avoidance predates its analysis in other more recent terms within associative learning. Nevertheless, if we do not keep apace with research developments in associative learning, we run the risk that the RFT analysis of fear and avoidance will be marginalized in the literature on this topic, and its impact diminished. By adapting existing experimental paradigms such as affective priming (e.g., Hermans, Vansteenwegen, Crombez, Baeyens, & Eelen, 2002) and approach-avoidance conflict tasks (De Houwer, Crombez, Baeyens, & Hermans, 2001) for use in RFT research, contact can be maintained and cross-disciplinary collaboration fostered. Another promising opportunity for collaboration with those in the associative learning tradition in the analysis of fear and avoidance relates to the study of various "pathways" by which avoidance may emerge.

Pathways to Fear and Avoidance

Fear acquired through pathways other than Pavlovian conditioning, such as verbal instructions and social observation, is often indistinguishable from directly learned fear (Olsson & Phelps, 2004, 2007). Further analysis of fear and avoidance acquired via indirect pathways may provide a novel basis for understanding the etiology of anxiety disorders (Askew & Field, 2008). It remains to be seen, however, whether avoidance rates and expectancies of an aversive stimulus (i.e., threat beliefs) acquired via symbolic generalization are comparable to those acquired via other pathways, such as instructions, and to what extent they differ, if at all, from directly learned avoidance.

Dymond, Schlund, Roche, De Houwer and Freegard (2012) sought to examine whether learned, instructed, and derived pathways result in equivalent levels of avoidance and outcome ratings. Following fear conditioning in which one stimulus was paired with shock (CS+) and another was not (CS-), participants either learned or were instructed to make an avoidance response that canceled impending shock. The three groups were then tested with learned CS+ and CS- (learned group), instructed CS+ (instructed group), and derived CS+ (derived group) presentations. Results showed similar levels of avoidance behavior and threat-belief ratings across the pathways despite the different ways in which they were acquired. These findings show, for the first time, the acquisition and maintenance of equivalent levels of avoidance acquired via learned, instructed and derived pathways. As such, they add to the existing literature on alternative pathways to fear and avoidance (Askew & Field, 2008; Dymond & Roche, 2009) and provide support for the RFT model of psychopathology in terms of derived relational processes.

Specific Phobia

According to RFT, phobic avoidance occurs in the presence of a wide range of stimuli and situations based on the actual and, more often than not, inferred presence of the aversive event. In this way, verbal relations may mediate the acquisition and generalization of fear and avoidance along indirect pathways (Dymond & Roche, 2009; Dymond et al., 2011). If this assumption is correct, then groups that already differ on the basis of specific self-reported traits may also show different levels of derived avoidance.

In one study, Smyth et al. (2006, Experiment 2) sought to examine whether spider-fearful and non-spider-fearful participants would show differing levels of derived transfer of self-reported arousal functions. The authors first trained and tested for the formation of 2, three-member equivalence relations (A1-B1-C1 and A2-B2-C2) before training (with A1 and A2) and testing (with C1 and C2) for a transfer of simple discrimination functions. Next, a stimulus pairing observation procedure (SPOP) was used to establish conditional relations A1-B1, A2-B2, A1-C1 and A2-C2. During a block of video-pairing trials, A1 was paired with spider attack scenes from a well-known movie while A2 was paired with a blank screen. Smyth et al. tested for transfer of self-reported arousal functions by asking participants to categorize unseen videos labeled with the remaining members of the equivalence relations and found that spider-fearful participants showed significantly greater arousal than non-fearful participants to direct and indirectly related stimuli. The findings of Smyth et al. (2006) provide support for the RFT and derived relations approach to the acquisition of anxiety and, in this case, specific phobia.

A recent study by Dymond, Schlund, Roche, and Whelan (in press) sought to replicate and extend these findings to an examination of derived avoidance with a between-groups design. High and low spider fearful participants learned relations between arbitrary visual cues resulting in two stimulus equivalence relations (A=B=C; X=Y=Z). Next, one cue (B) was established as a CS+ that signaled onset of spider images and prompted avoidance (threat cue), and another cue (Y) was established as a CS- that signaled the absence of such images and did not prompt avoidance (safety cue). High spider fearful individuals exhibited greater levels of derived threat-avoidance to cues indirectly related to the threat cue (A and C) than low spider fearful individuals, and both groups showed non-avoidance to cues indirectly related to the safety cue (X and Z). These findings showed, for the first time, differential levels of clinically relevant avoidance as a function of self-reported phobia status and have implications for the RFT understanding of the acquisition of excessive avoidance.

The Analysis of Thought Suppression Coping Strategies

ACT proposes that common-sense coping strategies in stressful situations, such as experiential avoidance, including thought suppression, may not only not alleviate suffering but may even exacerbate it. Indeed, copious evidence now exists that this is likely the case (Hayes et al., 2006). However, laboratory controlled studies of thought suppression in terms of controlled relational process are rare. One exception is a study by Hooper, Saunders and McHugh (2010) that involved first establishing 3, three-member equivalence relations. Next, participants were instructed to suppress all thoughts of a target word that was a member of one of the equivalence relations and could escape words presented to them on a computer screen. During the crucial test, participants escaped not only the to-be-suppressed stimulus but also stimuli indirectly related (in this case, via symmetry and transitivity). This demonstrated that ineffective coping strategies themselves can generalize across stimuli to an increasingly wider array of stimuli, thereby potentially exacerbating the original problem. These findings justify replication and extension to multiple stimulus relations and to a comparison of potentially differential derived outcomes in subclinical and clinical populations for whom thought suppression is particularly problematic.

Laboratory Analogs of Therapeutic Processes

Numerous scientific constructs have been offered by the ACT literature to guide clinicians and researchers, including "cognitive defusion", which is likely the most widely discussed and easily experimentally addressed concept within the ACT literature. "Middle-level terms" such as cognitive defusion represent classes of behavioral events functionally based on RFT and basic behavioral principles (Vilardaga, Hayes, Levin, & Muto, 2009). Nevertheless, it is crucial that use of these terms is based on experimental verification of the assumed underlying processes because there is a paucity of process-level analyses elucidating the underlying mechanisms on which such concepts are based. In this final section, we describe two possible therapy analogs of cognitive defusion as a therapeutic process.

Cognitive defusion techniques manipulate the context in which aversive verbal events occur in order to de-fuse or attenuate their literal meaning and reduce their subsequent control over behavior (see Blackledge & Drake, this volume). From an RFT perspective, defusion produces meaningful behavior change through the disruption of problematic derived stimulus functions by introducing alternative contextual cues that displace the cues supporting the transformation of those functions (see Blackledge, 2007). While the clinical efficacy of defusion protocols is not in question (e.g., Bach & Hayes, 2002; Masuda et al., 2004), its supposed mechanisms have yet to be empirically verified. There is simply a lack of empirical evidence to indicate, firstly, that derived avoidance and anxiety functions can be brought under contextual control and, secondly, that this might reduce maladaptive responding in certain contexts. Indeed, one study by Roche et al. (2008) assessed whether derived extinction may in fact be all that is involved in defusion.

In that study, Roche et al. (2008) compared the relative merits of direct vs. derived extinction of avoidance in an attempt to ascertain whether derived extinction was at least as powerful as direct extinction of aversive stimulus functions. This would help to clarify whether or not defusion exercises should even, in principle, achieve the same or superior effects as traditional exposure techniques used to deal with directly learned fear and avoidance. The authors exposed participants to a relational training and testing sequence identical to that employed by Dymond et al. (2007), which resulted in the following derived Same and Opposite relations: Same B1-C1, Same B2-C2, Opposite B1-C2 and Opposite B1-C1. Next, participants were exposed to a signaled avoidance conditioning procedure in which B1 served as a discriminative stimulus for an avoidance response that canceled upcoming aversive pictures, and B2 served as a discriminative stimulus for non-avoidance, respectively. All participants who showed stable avoidance responding to B1 also showed derived avoidance responses to C1, but not to C2, thus replicating the findings of Dymond et al. (2007, 2008).

In a subsequent phase, participants were exposed to either the direct or derived extinction of avoidance procedure in which avoidance was no longer effective and aversive images followed B1 and B2 (direct) or C1 and C2 (derived). Participants exposed to direct extinction showed derived extinction of avoidance to C1, whereas subjects exposed to derived extinction showed derived extinction of avoidance to B1. Extinction of the derived avoidance functions of C1 was found to be more effective than extinction of the directly established avoidance functions of B1. This suggests it is relatively easy to extinguish avoidance responses to stimuli that have acquired their functions through derived relational processes, whereas avoidance responses to directly aversive stimuli are more resistant to extinction. Moreover, derived extinction was more effective than direct extinction at extinguishing avoidance to all members of the network, with extinction of avoidance to C1 transferring readily to B1 for most participants in the derived extinction condition, whereas extinction of of B1 in the direct extinction condition did not transfer to C1. In simple terms, the effect of the extinction procedure generalized across the derived relations more effectively when the derived, rather than the directly conditioned, aversive stimulus was targeted. Even more surprisingly, a greater extinction of avoidance was observed for B1 when C1 was targeted than was observed for B1 when B1 itself was targeted. In any case, the fact that derived extinction was demonstrated at all, and at least equaled the effects of direct extinction, supports the idea that defusion may in principle be effective as a treatment for avoidance if it does indeed involve the derived transfer of extinction.

It is still unclear whether the efficacy of defusion lies in its facilitation of (derived) exposure to aversive verbal stimuli in multiple contexts (e.g., Arch & Craske, 2008; Roche et al., 2008; Vansteenwegen et al., 2007) or merely in just one therapeutic context. This possibility could be tested as follows. First, three-member equivalence relations could be established such that a stimulus C is equivalent to stimulus A. Second, in a specific context such as an orange background, a controlled and quantifiable avoidance response (e.g., a key press) to C may be negatively reinforced through the contingent removal of an aversive stimulus (e.g.,

Dymond et al., 2007, 2011). To replicate defusion protocols, additional responses to C may now be reinforced in new contexts. For instance, when the background is green, hand-waving in the presence of C might be reinforced; when the background is red, clapping hands in the presence of C might be reinforced; when the background is blue, tapping feet in the presence of C might be reinforced, and so on. Finally, the crucial test for the "therapeutic" benefits of this experimental manipulation would involve testing for derived avoidance of stimulus A both in a novel context, such as a pink background color, and in the original conditioning contexts.

It might be predicted that the derived avoidance of stimulus A would be determined by the contextual cue and, thus, levels of derived avoidance would be lower in the novel relative to the original context. Of course, this possibility remains to be investigated, but by demonstrating that the meaning of an aversive thought is contextually determined, a client may show less anxiety and avoidance to that thought and its referent events in post-therapeutic situations. Highly controlled RFT-inspired experimental paradigms such as this are similar to experimental psychopathology approaches (e.g., Vansteenwegen, et al., 2007) and can greatly enhance conceptual validity and clarity of cognitive defusion and other ACT based components.

Conclusions

We have illustrated here that since the original "purple RFT book", some ground has been made in the conceptual and empirical analysis of derived relational processes as they relate to psychopathology, and in particular avoidance and anxiety. However, there is no denying that much more research needs to be done in the laboratory in order to prevent our literature base becoming too top-heavy with interpretive and applied analyses. An imbalance of this kind would not be in keeping with the basic-applied research focus of RFT and contextual behavioral science more generally (see Vilardaga et al., 2009). Specifically, more research is needed with both subclinical and non-clinical samples. Perhaps more importantly, however, we need process-level studies that both demonstrate and control the emergence and reduction of clinically relevant behaviors in the laboratory. No amount of applied research in the therapy setting can satisfy the scientific vision of RFT if more basic research is not conducted.

One exciting new avenue of research and collaboration in which RFT researchers should aim to participate is the neuroimaging of psychopathology processes. Not only do such studies build bridges to neighboring sciences, but studies involving neuroimaging techniques have high visibility in the scientific literature. Given our uniquely precise and progressive approach to psychopathology we can easily pique the interest of other scientists who may have the techniques and instrumentation to map brain processes onto behavioral processes, but who have no particular commitment to any one behavioral processes other than for their novelty. Instances of such collaborations have already begun and look set to continue (e.g., Hinton, Dymond, von Hecker, & Evans, 2010; Schlund et al., 2010; Schlund, Magee, & Hudgins, 2011; and see, Whelan and Schlund, this volume).

RFT researchers are well equipped conceptually and methodologically to address many issues relevant to the emergence, maintenance and treatment of psychopathology. The field is wide open for research that is based on the functional-analytic approach (see De Houwer, 2011). Indeed, the time is ripe for a major leap forward in the RFT analysis of psychopathology that will require re-focusing on the importance of basic laboratory research on which, ultimately, the successful application of RFT may stand or fall.

References

Arch, J., & Craske, M.G. (2008). ACT and CBT for anxiety disorders: different treatments, similar mechanisms? *Clinical Psychology: Science and Practice, 15,* 263-279

Askew, C., & Field, A. P. (2008). The vicarious learning pathway to fear 40 years on. *Clinical Psychology Review, 28,*1249-1265.

Augustson, E. M., & Dougher, M. J. (1997). The transfer of avoidance evoking functions through stimulus equivalence classes. *Journal of Behavior Therapy & Experimental Psychiatry, 28,* 181-191.

Bach, P., & Hayes, S. C. (2002). The use of acceptance and commitment therapy to prevent the rehospitalization of psychotic patients: A randomized controlled trial. *Journal of Consulting and Clinical Psychology, 70,* 1129-1139.

Beckers, T., Krypotos, A-M., Boddez, Y., Effting, M., & Kindt, M. (2013). What's wrong with fear conditioning? *Biological Psychology, 92,* 90-96.

Bentall, R. P. (2006). Madness explained: Why we must reject the Kraepelinian paradigm and replace it with a 'complaint-orientated' approach to understanding mental illness. *Medical Hypotheses 66,* 220-233.

Blackledge, J. T., & Barnes-Holmes, D. (2009). Core processes in acceptance and commitment therapy. In J. T. Blackledge, J. Ciarrochi, & F. P. Deane (Eds.), *Acceptance and Commitment Therapy: Contemporary theory research and practice* (pp. 1-40). Sydney: Australian Academic Press.

Bond, F. W., Hayes, S. C., Baer, R. A., Carpenter, K. C., Guenole, N., Orcutt, H. K., Waltz, T. and Zettle, R. D. (2011). Preliminary psychometric properties of the Acceptance and Action Questionnaire—II: A revised measure of psychological flexibility and acceptance. *Behavior Therapy, 42,* 676-688.

Cochrane, A., Barnes-Holmes, D., & Barnes-Holmes, Y. (2008). The Perceived Threat Behavioural Approach Test (PT-BAT): Measuring avoidance in high-, mid-, and low- spider fearful participants. *The Psychological Record, 58,* 585–596.

De Houwer, J. (2011). Why the cognitive approach in psychology would profit from a functional approach and vice versa. *Perspectives on Psychological Science, 6,* 202-209.

De Houwer, J., Crombez, G., Baeyens, F., & Hermans, D. (2001). On the generality of the affective Simon effect. *Cognition and Emotion, 15,* 189-206.

Dougher, M. J., Augustson, E., Markham, M. R., Greenway, D. E., & Wulfert, E. (1994). The transfer of respondent eliciting and extinction functions through stimulus equivalence classes. *Journal of the Experimental Analysis of Behavior, 62,* 331-351.

Dougher, M. J., Hamilton, D. A., Fink, B. C., & Harrington, J. (2007). Transformation of the discriminative and eliciting functions of generalized relational stimuli. *Journal of the Experimental Analysis of Behavior, 88,* 179-197.

Dunsmoor, J. E., Martin, A., & LaBar, K. S. (2012). Role of conceptual knowledge in learning and retention of conditioned fear. *Biological Psychology, 89*, 300-305.

Dunsmoor, J. E., Mitroff, S. R., & LaBar, K. S. (2009). Generalization of conditioned fear along a dimension of increasing fear intensity. *Learning & Memory, 16*, 460-469.

Dunsmoor, J. E., White, A. J., & LaBar, K. S. (2011). Conceptual similarity promotes generalization of higher order fear learning. *Learning & Memory, 18*, 156-160.

Dymond, S., & Barnes, D. (1997). Behavior-analytic approaches to self-awareness. *The Psychological Record, 47*, 181-200.

Dymond, S., May, R. J., Munnelly, A., & Hoon, A. E. (2010). Evaluating the evidence base for relational frame theory: A citation analysis. *The Behavior Analyst, 33*, 97-117.

Dymond, S. & Rehfeldt, R. A. (2000). Understanding complex behavior: The transformation of stimulus functions. *The Behavior Analyst, 23*, 239-254.

Dymond, S., & Roche, B. (2009). A contemporary behavior analysis of anxiety and avoidance. *The Behavior Analyst, 32*, 7-28.

Dymond, S., Roche, B., Forsyth, J. P., Whelan, R., & Rhoden, J. (2007). Transformation of avoidance response functions in accordance with the relational frames of same and opposite. *Journal of the Experimental Analysis of Behavior, 88*, 249-262.

Dymond, S., Roche, B., Forsyth, J. P., Whelan, R., & Rhoden, J. (2008). Derived avoidance learning: transformation of avoidance response functions in accordance with the relational frames of same and opposite. *The Psychological Record, 58*, 271-288.

Dymond, S., Schlund, M. W., Roche, B., De Houwer, J., & Freegard, G. (2012). Safe from harm: Learned, instructed, and symbolic generalization pathways of human threat-avoidance. *PLoS ONE, 7*(10): e47539. doi:10.1371/journal.pone.0047539.

Dymond, S., Schlund, M. W., Roche, B., & Whelan, R. (in press). The spread of fear: Symbolic generalization mediates graded threat-avoidance in specific phobia. *Journal of Psychiatry Research.*

Dymond, S., Schlund, M., Roche, B., Whelan, R., Richards, J., & Davies, C. (2011). Inferred threat and safety: Symbolic generalization of human avoidance learning. *Behaviour Research & Therapy, 49*, 614-621.

Forsyth, J. P., Eifert, G. H., & Barrios, V. (2006). Fear conditioning in an emotion regulation context: A fresh perspective on the origins of anxiety disorders. In M. G. Craske, D. Hermans, & D. Vansteenwegen (Eds.), *Fear and learning: From basic processes to clinical applications* (pp. 133-153). Washington, DC: American Psychological Association.

Gannon, S., Roche, B., Kanter, J., Forsyth, J. P., & Linehan, C. (2011). A derived relations analysis of approach-avoidance conflict: Implications for the behavioural analysis of human anxiety. *The Psychological Record, 61*, 227-252.

Gutiérrez, O., Luciano, C., Rodríguez, M., & Fink, B. C. (2004). Comparison between an acceptance-based and a cognitive-control-based protocol for coping with pain. *Behavior Therapy, 35,* 767–783.

Hayes, S. C. (1976). The role of approach contingencies in avoidance behavior. *Behavior Therapy, 7,* 28–36.

Hayes, S. C., Barnes-Holmes, D., & Roche, B. (2001). *Relational frame theory: A post-Skinnerian account of human language and cognition.* New York: Kluwer Academic.

Hayes, S. C., Luoma, J. B., Bond, F. W., Masuda, A., & Lillis, J. (2006). Acceptance and commitment therapy: Model, processes, and outcomes. *Behaviour Research & Therapy, 44,* 1–25.

Helzer, J. E., Kraemer, H. C., & Krueger, R. F. (2006). The feasibility and need for dimensional psychiatric diagnoses. *Psychological Medicine, 36,* 1671-80.

Hermans, D., Vansteenwegen, D., Crombez, G., Baeyens, F., & Eelen, P. (2002) Expectancy-learning and evaluative learning in human classical conditioning: Affective priming as an indirect and unobtrusive measure of conditioned stimulus valence. *Behaviour Research & Therapy, 40,* 217-234.

Hinton, E. C., Dymond, S., von Hecker, U., & Evans, C. J. (2010). Neural correlates of relational reasoning and the symbolic distance effect: Involvement of parietal cortex. *Neuroscience, 168,* 138-148. Corrigendum, *170,* 1345.

Hooper, N., Saunders, J., & McHugh, L. (2010). The generalization of thought suppression. *Learning and Behavior, 38,* 160-168.

Keogh, E., Bond, F. W., Hanmer, R., & Tilston, J. (2005). Comparing acceptance and control-based coping instructions on the cold-pressor pain experiences of healthy men and women. *European Journal of Pain, 9,* 591-598.

Lissek, S., Rabin, S., Heller, R. E., Lukenbaugh, D., Geraci, M., Pine, D. S., & Grillon, C. (2010). Overgeneralization of conditioned fear as a pathogenic marker of panic disorder. *American Journal of Psychiatry, 167,* 47-55.

Lommen, M., Engelhard, I., & van den Hout, M. (2010). Neuroticism and avoidance of ambiguous stimuli: Better safe than sorry? *Personality & Individual Differences, 49,* 1001-1006.

Lovibond, P. F. (2006). Fear and avoidance: an integrated expectancy model. In M. G. Craske, D. Hermans, & D. Vansteenwegen (Eds.), *Fear and learning: From basic processes to clinical applications* (pp. 117-132). Washington, DC: American Psychological Association.

Masuda, A., Hayes, S. C., Sackett, C. F., & Twohig, M. P. (2004). Cognitive defusion and self-relevant negative thoughts: Examining the impact of a ninety-year-old technique. *Behaviour Research and Therapy, 42,* 477-485.

McMullen, J., Barnes-Holmes, D., Barnes-Holmes, Y., Stewart, I., & Cochrane, A. (2007) Acceptance versus distraction: Brief instructions, metaphors, and exercises in increasing tolerance for self-delivered electric shocks. *Behaviour Research and Therapy, 46,* 122-129.

Muris, P., & Field, A. P. (2010). The role of verbal threat information in the development of childhood fear. "Beware the Jabberwock!" *Clinical Child & Family Psychology Review, 13*, 129-150.

Olsson, A., Phelps, E. A. (2004). Learned fear of "unseen" faces after Pavlovian, observational, and instructed fear. *Psychological Science, 15*, 822–828.

Olsson, A., & Phelps, E. A. (2007). Social learning of fear. *Nature Neuroscience, 10*, 1095-1102.

Roche, B., Barnes-Holmes, D., Smeets, P. M., Barnes-Holmes, Y. & McGeady, S. (2000). Contextual control over the derived transformation of discriminative and sexual arousal functions. *The Psychological Record, 50*, 267-291.

Roche, B., & Barnes, D. (1997). A transformation of respondently conditioned stimulus function in accordance with arbitrarily applicable relations. *Journal of the Experimental Analysis of Behavior, 67*, 275-301.

Roche, B., Forsyth, J., & Maher, E. (2007). The impact of demand characteristics on brief acceptance and control-based interventions for pain tolerance. *Cognitive and Behavioral Practice, 14*, 381-393.

Roche, B., Kanter, J. W., Brown, K. R., Dymond, S., & Fogarty, C. C. (2008). A comparison of 'direct' versus 'derived' extinction of avoidance. *The Psychological Record, 58*, 443-464.

Rodriguez Valverde, M., Luciano, C., & Barnes-Holmes, D. (2009). Transfer of aversive respondent elicitation in accordance with equivalence relations. *Journal of the Experimental Analysis of Behavior, 92*, 85-111.

Schlund, M. W., Magee, S., & Hudgins, C. D. (2011). Human avoidance and approach learning: Evidence for overlapping neural systems and experiential avoidance modulation of avoidance neurocircuitry. *Behavioural Brain Research, 225*, 437-448.

Schlund, M. W., Siegle, G. J., Ladouceur, C. D., Silk, J. S., Cataldo, M. F., Forbes, E. E., Dahl, R. E., & Ryan, N. D. (2010). Nothing to fear? Neural systems supporting avoidance behavior in healthy youths. *NeuroImage, 52*, 710-719.

Smyth, S., Barnes-Holmes, D., & Forsyth, J. P. (2006). Derived transfer of self-reported arousal functions. *Journal of the Experimental Analysis of Behavior, 85*, 223-246.

Stewart, I., & McElwee, J. (2009). Relational responding and conditional discrimination procedures: An apparent inconsistency and clarification. *The Behavior Analyst, 32*, 309-317.

van Os, J., Linscott, R. J., Myin-Germeys, I., Delespaul, P., & Krabbendam, L. (2009). A systematic review and meta-analysis of the psychosis continuum: Evidence for a psychosis proneness-persistence-impairment model of psychotic disorder. *Psychological Medicine, 39*, 179-195.

Vansteenwegen, D., Vervliet, B., Iberico, C., Baeyens, F., Van den Bergh, O., & Hermans, D. (2007). The repeated confrontation with videotapes of spiders in multiple contexts attenuates renewal of fear in spider-anxious students. *Behaviour Research and Therapy, 45*, 1169-1179.

Vervliet, B., Kindt, K., Vansteenwegen, D., & Hermans, D. (2010). Fear generalization in humans: Impact of prior non-fearful experiences. *Behaviour Research & Therapy, 48,* 1078-1084.

Vilardaga, R., Hayes, S. C., Levin, M. E., & Muto, T. (2009). Creating a strategy for progress: A contextual behavioral science approach. *The Behavior Analyst, 32,* 105–133.

Wagener, A. L., & Zettle, R. D. (2011). Targeting fear of spiders with control-, acceptance-, and information-based approaches. *The Psychological Record, 61,* 77–92.

Wetterneck, C. T., Burgess, A. J., Short, M. B., Smith, A. H., Cervantes, M. E. (2012). The role of sexual compulsivity, impulsivity and experiential avoidance in internet pornography use. *The Psychological Record, 62.1,* 3-17.

Zettle, R. D., Petersen, C. L., Hocker, T. A., & Provines, J. L. (2007). Responding to a challenging perceptual-motor task as a function of level of experiential avoidance. *The Psychological Record, 57,* 49–62.

CHAPTER TEN

Acceptance and Commitment Therapy: Empirical and Theoretical Considerations

John T. Blackledge

Morehead State University

Chad E. Drake

Southern Illinois University

Acceptance and Commitment Therapy (ACT; see, for example, Hayes, Strosahl, & Wilson, 1999) is a contemporary form of behavior therapy informed by Relational Frame Theory (RFT), functional contextualism (e.g., Gifford & Hayes, 1999), and aspects of radical behaviorism (e.g., Skinner, 1953). This chapter will provide a brief sketch of ACT's origins, underlying theory, empirical evidence, links to RFT research and theory, and future research challenges.

ACT from a Historical Perspective: The Cognitive Revolution and ACT

The cognitive revolution in psychology occurred over a roughly 15-year period (beginning around 1955) primarily as a reaction against behavioral theory. In discussing the origins of the cognitive revolution and its eventual relationship to ACT, a distinction must be made between the cognitive *science revolution* and the *evolution* of cognitive *therapy*. Both aspects of the cognitive revolution, and its influence on the development of ACT, will be discussed in turn.

ACT and the Cognitive Science Revolution

Methodological and radical behaviorism had a profound and direct impact on the origins of the cognitive *science* revolution. A driving force behind the cognitive science revolution involved behaviorism's apparent inability (and lack of intent) to account for cognitive processes and their effects on human behavior. Skinner explicitly rejected that thoughts could play a causal role in determining behavior, referring to verbal behavior as "dependent variables" (Skinner, 1992, p. 14). The emerging cadre of cognitive scientists rejected Skinner's perspective, insisting that human cognition was much more complex and generative than could be accounted for by direct operant conditioning alone (e.g., Chomsky, 1957) and that cognition readily causes human behavior (e.g., Clark & Chalmers, 1998). Chomsky's scathing review of *Verbal Behavior* arguably served as a catalyst for many psychologists to reject behaviorism in its entirety, and cognitive psychology proved to be an attractive alternative (see Hatfield, 2002, for a more in-depth discussion of the transition from behavioral to cognitive psychology).

Philosophical differences from radical behaviorism became evident as well, with cognitive scientists largely adopting a mechanistic perspective (see Stepp, Chemero & Turvey, 2011; see also Pepper, 1961) where the primary analytic goal is to thoroughly delineate a model of cognition that allows an increased ability to predict behavior. This stood in contrast to Skinner's contextual assumption that the primary purpose of a science of psychology should be to enhance the prediction and control of human behavior (e.g., Skinner, 1953). These philosophical differences go far in explaining why cognitive therapy is minimally informed by cognitive science, and behavior therapy is relatively closely linked to behavioral science. Placing emphasis on the creation and testing of a cognitive model that enhances prediction of human behavior does not necessarily translate to the creation of interventions designed to change (or control) human behavior. Behaviorism's typical focus on enhancing the prediction *and* control of behavior tends to lead basic researchers to more readily consider the practical applications of their work and applied researchers to more carefully consider what basic researchers have found under more tightly controlled conditions.

ACT and the Cognitive Therapy Evolution

While aspects of behavioral science had a clear and direct impact on the origins and rise of cognitive science, the transition from radical behavior therapy appears to have had less of a reactionary impact on the development of cognitive therapy. The cognitive therapies of Albert Ellis and Aaron Beck have been described more as reactions against psychodynamic therapy. Ellis was trained as a psychoanalyst, and created his original Rational Therapy in part as an antidote to the slow pace of psychoanalysis (Dowd, 2004); Beck created Cognitive Therapy partly as a reaction to the lack of empirical evidence for psychodynamic therapy (Hollon, 2010). Beck freely acknowledges incorporating many behavioral techniques into his brand of cognitive therapy (e.g., Beck, Rush, Shaw, & Emery, 1979), and Ellis eventually changed the name of his treatment to Rational Emotive Cognitive Behavior Therapy. Indeed, two other noted cognitive therapy pioneers, Albert Bandura and Donald Meichenbaum, developed treatments that attempted to capitalize on the role one's internal dialogue appeared to play in influencing behavior, but conceptualized these treatments essentially in behavioral terms (Dowd, 2004). Thus, the shift from pure behavior therapy to cognitive therapy largely involved integrating newfound cognitive techniques with standard behavioral ones, as the phrase cognitive-behavior therapy would suggest. But clearly, given the emphasis placed on cognitive change techniques by both Ellis and Beck, both found behavior therapy lacking. Both emphasized the often dramatic role that irrational beliefs (Dryden & Ellis, 1986) or cognitive distortions (Beck, 1967) play in driving dysfunctional behavior, and both call explicitly for the use of an arsenal of cognitive change techniques.

For present purposes, it is worth noting that cognitive therapy is at best minimally informed by the results of cognitive science research (see, for example, Tataryn, Nadel, & Jacobs, 1989). The asynchronous trajectories of cognitive science and cognitive therapy serve as a counterpoint to a primary goal of the contextual behavior science movement. From the beginning, ACT (then known as comprehensive distancing; Zettle, 2005) was intended to map onto basic behavioral principles, especially the emerging behavioral principles behind relational frame theory, and to develop in response to relevant empirical findings in RFT research (Zettle, 2005). This is not to say that ACT is thoroughly grounded in RFT or behavioral theory writ large. Much work remains in creating a seamless integration between ACT and RFT, though details regarding much of the empirical and theoretical overlap between the two entities are discussed later in this chapter.

ACT: Revolution and Evolution

So, the question remains: How did the cognitive science revolution and cognitive therapy evolution impact the creation of ACT? Spurred in part by the relative empirical success of cognitive therapy (and to a broader degree, cognitive-behavioral therapy) but concerned with the theoretical issues regarding the cognitive model previously discussed, Steven Hayes and colleagues began attempting to conceptualize cognitive techniques from a conventional behavioral perspective beginning in the mid-late 1970s (Zettle, 2005). This culminated in the publication of Zettle and Hayes (1982), an attempt to conceptualize various cognitive restructuring techniques advocated by Aaron Beck and Albert Ellis in behavioral terms. One of the cognitive techniques conceptualized in that chapter was Beck's "cognitive distancing" (Hollon & Beck, 1979, p.189), a necessary first step in the process of cognitive restructuring where a client is taught to notice problematic automatic thoughts and core beliefs and view them as hypotheses rather than established facts. Hayes and colleagues evidently viewed an expanded and more pervasive form of cognitive distancing as a necessary and sufficient means of producing therapeutic change, theorizing that simple recognition of maladaptive behavioral rules and inaccurate descriptions as thoughts, rather than absolute truths, could be sufficient to produce change. "Comprehensive distancing" (Zettle, 2005, p. 80), as it was called, continued to develop over the next decades, and with the addition of explicit values and self-as-context aspects (among others) came to be known as ACT (see Hayes, Strosahl & Wilson, 2011 and Hayes & Strosahl, 2004, for expanded treatments of ACT).

The initial behavioral rationale behind comprehensive distancing was purely Skinnerian. Simply put, particular contexts support the creation of verbal rules and drive behavior that is consistent with those rules. Comprehensive distancing strategies create a context where behavior inconsistent with those rules becomes more probable. However, the deeper into the analysis Hayes and colleagues went, the more they realized that Skinner's (1969) definition of verbal rules as contingency specifying stimuli was inadequate for the task. Hayes (1987) provided an analysis of rule-governed behavior that was clearly distinguishable from Skinner's analysis and that paved the way for the development of RFT. The resulting analysis ultimately led to a conclusion similar to the one that helped spur the cognitive revolution: existing behavioral principles (circa 1957-1986) could not adequately conceptualize human language and cognition, and therapeutic techniques that could address language and cognition were needed. But while the path taken by cognitive theorists and therapists sharply diverged from behavioral theory, the ACT path was marked by the parsimonious addition of RFT and the preservation of behaviorism's focus on manipulable contextual variables and pursuit of enhanced prediction and control at both the basic and applied experimental levels. Thus, the development of ACT and RFT was in some ways a reaction both to the cognitive (r)evolution and conventional behavioral theory. We now move to a description of the core components of ACT.

Core Principles and Techniques of ACT

Dysfunctional Rule-Governance

In addition to the advances and positive effects in human activity based on verbal behavior, the ability to derive and transform stimulus functions adds a new level of possibilities in respect to behavioral problems. A verbal human can spend a lifetime responding to and working for events primarily on the basis of their derived functions rather than the direct stimulation available in the environment. As such, pervasive patterns of escape from and avoidance of particular verbal experiences or a rigidly persistent pursuit or maintenance of alternative verbal experiences may translate into an array of psychological problems (Hayes, Wilson, Gifford, Follette, & Strosahl, 1996).

When dysfunctional behavior is generated and maintained on the basis of verbal stimulation, it may be regarded as problematic rule-governance. Hayes, Kohlenberg, and Melancon (1989) have discussed ways in which rules may exert dysfunctional influence. In some cases, appropriate rules may not be adequately formulated or followed. In others, the rules may themselves be dysfunctional or adherence to them may be excessive. This latter case presents a challenge for a clinician, as direct attempts to modify the rules or the consequences of following them may not effectively compete with the verbal contingencies that have maintained their power in the first place.

Preventing and Modifying Rule-Governance

ACT may be regarded as a treatment model for problematic rule-governance. In a sense, an ACT therapist may view clients as having problems in two basic respects: (1) their behavior is dominated by dysfunctional rules, a behavioral excess, and (2) they are failing to generate and respond to more useful rules, a behavioral deficit. Application of the ACT model involves training a collection of skills and promoting activities that in some cases disrupt rule-governance and in others enhance rule-governance. The common purpose of these activities is to increase psychological flexibility.

Psychological Flexibility

The notion of psychological flexibility first appeared in the ACT literature in Hayes, Strosahl, Bunting, Twohig, and Wilson (2004) and since then has been defined in multiple similar ways (e.g., Hayes et al., 2011). A recent review defines it as "the ability to contact consciously the present moment and the thoughts and feelings it contains more fully and without needless defense, and based on what the situation affords, to persist or change in behavior in the service of chosen values" (Hayes,

Villatte, Levin, & Hildebrandt, 2011, p. 155). This non-technical definition indicates that psychological flexibility refers to a class of behavior and is therefore subject to functional analysis, that it involves sensitivity to the details of one's direct experience (including private events), and that it allows either continuation or cessation of a response pattern based on personal evaluations of its utility. The construct reflects six separate but interwoven processes (Bach & Moran, 2008). Recently, these six processes have been clustered into three groupings of two similar processes each (Hayes et al., 2011). In the following sections, we will discuss the divisions of these midlevel (as opposed to technical) terms and provide prototypical examples of their delivery in treatment settings.

Self-Awareness and Perspective Taking

A fundamental feature of psychological flexibility involves training certain perspective taking repertoires that generate greater sensitivity to direct experience and an awareness of stable aspects of one's perspective over time. The two hexaflex processes are called self-as-process and self-as-context. Together, they are categorized as involving awareness.

Self-as-Process vs. Self-as-Product

Self-as-process involves purposeful attention to events that are happening here and now. Often referred to as present moment awareness, this awareness is inclusive of any event that is ongoing in the direct experience of the observer, in contrast to preoccupation with memories of the past, concerns about a hypothetical future, or events occurring in some other place. This awareness may involve sensory contact with any feature of the external environment (sights, smells, sounds, etc.) as well as the internal environment (sensations, emotions, thoughts). Engagement in this activity involves continual observation of these details as well as sensitivity to the inevitable moments when attention shifts away from these details and an opportunity to return to the present moment appears. Common present moment exercises in clinical settings involve meditative-like activities, such as performing a body scan of sensory details or imagining sitting beside a stream and writing each thought on a leaf and placing it into the stream. These exercises may be regarded as rehearsal of a skill set that can be engaged in at virtually any moment in life.

Self-as-Context vs. Self-as-Content

Self-as-context involves an awareness of the persistent, unchanging perspective from which one experiences all aspects of one's life. In other words, "self-as-context refers to the *I* who is always doing the discriminating" (Barnes-Holmes, Hayes, & Dymond, 2001, p. 129) of often rapidly changing sights, sounds, sensations, emotions, thoughts, memories, and so on. Given the change in context involved with viewing experiences from this perspective, problematic verbal rules and evaluative language begin to be experienced more as verbal events than definitive descriptions of reality. This stands in contrast to a sense of self-as-content, where the self is framed in coordination with one's verbal relations (e.g., as equivalent to one's negative self-evaluations, aversive emotions, and so on). Adopting a sense of self-as-content establishes a context where language functions literally and drives behavior. Thus, a central aim of ACT is to help clients adopt a sense of self-as-context when doing so is helpful. Invoking self-as-context may typically involve directing a client's attention to a variety of sensory, evocative, and cognitive experiences, and periodically asking the client to notice the consistent "you" common to these different experiences. A number of metaphors attempt to highlight the experience as well. The Chessboard Metaphor compares the internal conflict between positive and negative evaluations to the battle between the black and white pieces on a chessboard, where the observing self is like the board which contains this content without being affected or defined by it.

Self-as-process and self-as-context both emphasize the behavior of perceiving events from one's unique and private perspective. When adopting a sense of self-as-process, one becomes more aware of the moment-to-moment changes occurring in one's environment and thus better able to respond effectively to these changes. When adopting a sense of self-as-context, one can begin to respond in a more psychologically flexible manner to verbal events, especially when acceptance and defusion strategies are used in tandem.

Openness to Experience and Detachment from Literality

A second fundamental grouping of processes is designed to assist in the development of a different relationship with private experiences, particularly those that dysfunctionally connect to subsequent patterns of behavior. These two processes are defusion and acceptance. Together, they build a repertoire of openness.

Cognitive Defusion vs. Fusion

Defusion (originally called deliteralization; Hayes, Strosahl, & Wilson, 1999) involves distinguishing between the meaning of a thought and the direct sensory details of having that thought. Defusion, in a sense, "defuses" the potency found in the meaning of a particular cognitive experience. Whereas the literal meaning of a thought may drive a person to respond in a given manner, defusing from that thought provides the person with a sense of distance or detachment from the thought that reduces the necessity of a reaction. The thought may still occur, but the disruption or dilution of the meaning of that thought modifies its behavioral effects. Defusion may achieve its effects by displacing some of the key contextual features that support literal verbal transformations of function (see chapter 3 in this volume for a definition of this term), such as abiding by grammatical rules, speaking at a moderate pace (rather than much too fast or much too slow), or focusing on the content of language rather than the process of producing it (Blackledge, 2007). Perhaps the standard example of a defusion intervention is the "Milk" exercise, where a word is rapidly repeated for thirty seconds or more until the word loses literal meaning and only the sensory qualities (e.g., sound of the syllables, sensations of the mouth and throat) of the word remain.

Acceptance vs. Experiential Avoidance

Acceptance is the act of choosing unwanted or undesired elements of one's private experience and refraining from attempts to change or control them. Unlike tolerance or resignation, acceptance bears the qualities of attention, interest, curiosity, and willingness. Metaphorical language may be used to occasion acceptance, such as statements like "Can you create a space for this feeling?" and "What if you could make friends with this thought instead of fighting with it?" ACT protocols include multiple interventions designed to disrupt experiential avoidance. Generating "creative hopelessness" involves an assessment of the unworkability of control strategies over time, and "control is the problem" in some ways resembles a psychoeducational intervention that highlights the paradoxical effects of thought and emotion control strategies and orients the client toward acceptance and the enactment of personally-held values. Numerous metaphors highlighting acceptance vs. control are used, including the story of Joe the Bum, who arrives at your inclusive party and makes the event less pleasurable. Allowing Joe to remain at the party is necessary if one wants to fully experience the party—the alternative is to spend the evening preventing Joe from entering and removing him when he finds a way inside. Clients may connect their own experience in fighting with unwanted thoughts and feelings in the same way that they might encounter Joe knocking on their own door during a party.

Defusion and acceptance are both intended to modify a dysfunctional contingent relationship between private experience and behavior by changing the way one relates to those private experiences. With defusion, the literality of language products is disrupted. With acceptance, the evocative power of thoughts, feelings, and sensations is diminished.

Once disturbing thoughts and aversive emotions are experienced as events that can be safely experienced rather than necessarily avoided, such cognitive and affective events need not serve as barriers to effective value-driven action.

Motivation and Activation

The third pair of processes both embody the amplification of helpful sources of verbal influence into broader patterns of action. They are referred to as values and committed action.

Values vs. Control

Values clarification involves the exploration of and psychological contact with qualities of behavior that one finds personally meaningful, fulfilling, and purposeful. Values provide the motivation to continue living and to engage in the processes of treatment already discussed. They are not attainable as a goal and they are not a product of coercion by others or by an individual's private experiences. Endorsing and living in accordance with values is characterized as vital but not necessarily comfortable or pleasant at any given time. ACT therapists often relate values as direction one takes rather than a destination, highlighting that one can choose to follow a chosen trajectory at any given moment. A common values clarification exercise involves having a client imagine his or her own funeral, where attendees stand and make idealized statements about the client. Statements that provide information about the client's earnest cares and wants are then expanded upon to begin clarifying values in various domains of the client's life.

Committed Action vs. Inaction or Impulsivity

Committed action is the engagement in values-consistent activities. Whereas values provide the incentive for action, committed action involves both the moment-to-moment emission of values-consistent behavior and the pursuit of specific goals that one may encounter along a valued direction. These behaviors generally reflect a desire for outcomes that are attained over the long-term. In treatment settings, engaging in any of the previously detailed processes may qualify as committed action, as well as participation in more traditional interventions such as exposure, skills training, and behavioral activation. Outside of treatment, committed action may involve experiencing unpleasant or unwanted private experiences while engaging in values-consistent behavior. The Take Your Keys With You metaphor highlights this experience, where one's keys, regarded as painful thoughts and feelings that show up as a result of acting in a values-consistent manner, can be carried and used to "open up" a more vital, meaningful, and purposeful life.

Values and committed action build repertoires that allow more functional and life-enhancing values-based rules to govern behavior. Values clarification motivates and orients one toward the promise of a more workable and satisfying life. Committed action teaches people how to navigate their way through the challenges of a vital life.

Empirical Support for ACT

As an exhaustive and detailed literature review is beyond the scope of this chapter, readers are referred to Gaudiano (2011), Ost (2008; a meta-analysis of selected ACT outcome studies), Gaudiano (2009; a reply to that meta-analysis), and Levin and Hayes (2009b). At the present time, 51 randomized controlled trials investigating ACT's relative effectiveness have been published or are in press in peer-reviewed journals (see Table 1). All but two of these publications have occurred since 2000, and half since 2009 (after publishing two RCTs in the 1980s, Hayes and colleagues delayed conducting more so that the conceptual and theoretical underpinnings of ACT and RFT could be clarified; Zettle, 2005). As ACT was designed, in part, to impact psychological processes (e.g., experiential avoidance) presumed to be operative across a wide variety of psychological disorders, the populations targeted in these trials are diverse. At least four studies investigating ACT's effectiveness in treating depressive symptoms have been published (Forman, Herbert, Moitra, Yeomans, & Geller, 2007; Petersen & Zettle, 2009; Zettle & Hayes, 1986; Zettle & Rains, 1989), along with three studies testing ACT's effectiveness with psychosis. Seven studies comparing ACT to alternate therapies for chronic pain have been published (Dahl, Wilson & Nilsson, 2004; Johnston, Foster, Shennan, Starkey, & Johnson, 2010; Thorsell et al., 2011; Vowles et al., 2007; Wetherell et al., 2011; Wicksell, Melin, Lekander, & Olsson, 2009), with an additional seven studies involving ACT for the psychological complications associated with various medical diseases (Gregg, Callaghan, Hayes, & Glenn-Lawson, 2007; Lillis, Hayes, Bunting, & Masuda, 2009; Lundgren, Dahl, Melin, & Kees, 2006; Lundgren, Dahl, Yardi, & Melin, 2008; Paez, Luciano, & Gutierrez, 2007; Tapper et al., 2009; Westin et al., 2011). Studies investigating ACT's effect on a variety of nonclinical issues and processes have been published as well, including stigma (Hayes, Bissett, et al., 2004), therapist stress and burnout (Brinkborg, Michanek, Hesser, & Berglund, 2011), math anxiety (Zettle, 2003), worksite stress (Flaxman & Bond, 2010), and substance abuse counselors' willingness to implement evidence-based drug treatments (Varra, Hayes, Roget, & Fisher, 2008). Additionally, RCTs comparing ACT to alternate interventions for substance abusing (Hayes, Wilson, et al., 2004), borderline personality disordered (Gratz & Gunderson, 2006), and trichotillomanic (Woods, Wetterneck, & Flessner, 2006) clients have been published, as well as two RCTs investigating ACT with smoking cessation (Gifford, et al. 2004; Gifford et al., 2011). In virtually all cases, ACT was found to have superior effects relative to the comparison treatments (except in the case of math anxiety, where ACT was somewhat better than systematic desensitization for high experiential avoiders, but slightly worse for low experiential avoiders). It should be noted that a number of these studies had relatively small sample sizes, and that more research is needed to make more conclusive estimates of ACT's effects with these various populations. It should also be noted that a variety of single-subject, aggregated single-subject, and non-randomized studies on ACT have also been published.

Table 1: Published ACT Randomized Controlled Trials in Chronological Order

1986

Zettle, R. D. & Hayes, S. C. (1986). Dysfunctional control by client verbal behavior: The context of reason giving. *The Analysis of Verbal Behavior*, 4, 30-38.

1989

Zettle, R. D., & Rains, J. C. (1989). Group cognitive and contextual therapies in treatment of depression. *Journal of Clinical Psychology*, 45, 438-445.

2000

Bond, F. W. & Bunce, D. (2000). Mediators of change in emotion-focused and problem-focused worksite stress management interventions. *Journal of Occupational Health Psychology*, 5, 156-163.

2002

Bach, P. & Hayes, Steven C. (2002). The use of Acceptance and Commitment Therapy to prevent the rehospitalization of psychotic patients: A randomized controlled trial. *Journal of Consulting and Clinical Psychology*, 70 (5), 1129-1139.

2003

Zettle, R. D. (2003). Acceptance and commitment therapy (ACT) versus systematic desensitization in treatment of mathematics anxiety. *The Psychological Record*, 53, 197-215.

2004

Dahl, J., Wilson, K. G., & Nilsson, A. (2004). Acceptance and Commitment Therapy and the treatment of persons at risk for long-term disability resulting from stress and pain symptoms: A preliminary randomized trial. *Behavior Therapy*, 35, 785-802.

Gifford, E. V., Kohlenberg, B. S., Hayes, S. C., Antonuccio, D. O., Piasecki, M. M.., Rasmussen-Hall, M. L., & Palm, K. M. (2004). Acceptance theory-based treatment for smoking cessation: An initial trial of Acceptance and Commitment Therapy. *Behavior Therapy*, 35, 689-705.

Hayes, S. C., Wilson, K. G., Gifford, E. V., Bissett, R., Piasecki, M., Batten, S. V., Byrd, M., & Gregg, J. (2004). A randomized controlled trial of twelve-step facilitation and acceptance and commitment therapy with polysubstance abusing methadone maintained opiate addicts. *Behavior Therapy*, 35, 667-688.

Hayes, S. C., Bissett, R., Roget, N., Padilla, M., Kohlenberg, B. S., Fisher, G., Masuda, A., Pistorello, J., Rye, A. K., Berry, K. & Niccolls, R. (2004). The impact of acceptance and commitment training and multicultural training on the stigmatizing attitudes and professional burnout of substance abuse counselors. *Behavior Therapy*, 35, 821-835.

2006

Gaudiano, B.A., & Herbert, J.D. (2006). Acute treatment of inpatients with psychotic symptoms using Acceptance and Commitment Therapy. *Behaviour Research and Therapy*, 44, 415-437.

Gratz, K. L. & Gunderson, J. G. (2006). Preliminary data on an acceptance-based emotion regulation group intervention for deliberate self-harm among women with Borderline Personality Disorder. *Behavior Therapy*, 37, 25-35.

Lundgren, A. T., Dahl, J., Melin, L. & Kees, B. (2006). Evaluation of Acceptance and Commitment Therapy for drug refractory epilepsy: A randomized controlled trial in South Africa. *Epilepsia*, 47, 2173-2179.

Woods, D. W., Wetterneck, C. T., & Flessner, C. A. (2006) A controlled evaluation of Acceptance and Commitment Therapy plus habit reversal for trichotillomania. *Behaviour Research and Therapy*, 44, 639-656.

2007

Forman, E. M., Herbert, J. D., Moitra, E., Yeomans, P. D. & Geller, P. A. (2007). A randomized controlled effectiveness trial of Acceptance and Commitment Therapy and Cognitive Therapy for anxiety and depression. *Behavior Modification*, 31(6), 772-799.

Gregg, J. A., Callaghan, G. M., Hayes, S. C., & Glenn-Lawson, J. L. (2007). Improving diabetes self-management through acceptance, mindfulness, and values: A randomized controlled trial. *Journal of Consulting and Clinical Psychology*, 75(2), 336-343.

Lappalainen, R., Lehtonen, T., Skarp, E., Taubert, E., Ojanen, M., & Hayes, S. C. (2007). The impact of CBT and ACT models using psychology trainee therapists: A preliminary controlled effectiveness trial. *Behavior Modification*, 31, 488-511.

Masuda, A., Hayes, S. C., Fletcher, L. B., Seignourel, P. J., Bunting, K., Herbst, S. A., Twohig, M. P., & Lillis, J. (2007). The impact of Acceptance and Commitment Therapy versus education on stigma toward people with psychological disorders. *Behaviour Research and Therapy*, 45(11), 2764-2772.

Páez, M. B., Luciano, C., & Gutiérrez, O. (2007).Tratamiento psicológico para el afrontamiento del cáncer de mama. Estudio comparativo entre estrategias de aceptación y de control cognitivo. *Psicooncología*, 4, 75-95.

Vowles, K. E., McNeil, D. W., Gross, R. T. McDaniel, M. L., Mouse, A., Bates, M., Gallimore, P., & McCall, C. (2007). Effects of pain acceptance and pain control strategies on physical impairment in individuals with chronic low back pain. *Behavior Therapy*, 38, 412-425.

2008

Lundgren, T., Dahl, J., Yardi, N., & Melin, L. (2008). Acceptance and Commitment Therapy and yoga for drug-refractory epilepsy: A randomized controlled trial. *Epilepsy & Behavior*, 13, 102–108.

Luoma, J. B., Hayes, S. C., Roget, N., Fisher, G., Padilla, M., Bissett, R., Kohlenberg, B. K., Holt, C., & & Twohig, M. P. (2008). Augmenting continuing education with psychologically-focused group consultation: Effects on adoption of Group Drug Counseling. *Psychotherapy Theory, Research, Practice, Training*, 44, 463-469.

Roemer, L., Orsillo, S. M., & Salters-Pedneault, K. (2008). Efficacy of an acceptance-based behavior therapy for generalized anxiety disorder: Evaluation in a randomized controlled trial. *Journal of Consulting and Clinical Psychology*, 76, 1083-1089.

Varra, A. A., Hayes, S. C., Roget, N., & Fisher, G. (2008). A randomized control trial examining the effect of Acceptance and Commitment Training on clinician willingness to use evidence-based pharmacotherapy. *Journal of Consulting and Clinical Psychology*, 76, 449-458.

Wicksell, R, K., Ahlqvist, J., Bring, A., Melin, L. & Olsson, G. L. (2008). Can exposure and acceptance strategies improve functioning and quality of life in people with chronic pain and whiplash associated disorders (WAD): A randomized controlled trial. *Cognitive Behaviour Therapy*, 37, 1-14.

2009

Lillis, J., Hayes, S. C., Bunting, K., Masuda, A. (2009). Teaching acceptance and mindfulness to improve the lives of the obese: A preliminary test of a theoretical model. Annals of *Behavioral Medicine*, 37, 58-69.

Peterson, C. L. & Zettle, R. D. (2009). Treating inpatients with comorbid depression and alcohol use disorders: A comparisons of Acceptance and Commitment Therapy and treatment as usual. *The Psychological Record*, 59, 521-536.

Tapper, K., Shaw, C., Ilsley, J., Hill, A. J., Bond, F. W., & Moore, L. (2009). Exploratory randomised controlled trial of a mindfulness-based weight loss intervention for women. *Appetite*, 52, 396–404.

Wicksell, R. K., Melin, L., Lekander, M., & Olsson, G. L. (2009). Evaluating the effectiveness of exposure and acceptance strategies to improve functioning and quality of life in longstanding pediatric pain—A randomized controlled trial. *Pain*, 141, 248-257.

2010

Flaxman, P. E. & Bond, F. W. (2010). A randomised worksite comparison of acceptance and commitment therapy and stress inoculation training. *Behaviour Research and Therapy* 43, 816-820.

Flaxman, P. E., & Bond, F. W. (2010). Worksite stress management training: Moderated effects and clinical significance. *Journal of Occupational Health Psychology*, 15, 347-358.

Fledderus, M., Bohlmeijer, E. T., Smit, F., & Westerhof, G. J. (2010). Mental health promotion as a new goal in public mental health care: A randomized controlled trial of an intervention enhancing psychological flexibility. *American Journal of Public Health*, 10, 2372-2378.

Hinton, M. J. & Gaynor, S. T. (2010). Cognitive defusion for psychological distress, dysphoria, and low self-esteem: A randomized technique evaluation trial of vocalizing strategies. *International Journal of Behavioral Consultation and Therapy*, 6, 164-185.

Johnston, M., Foster, M., Shennan, J., Starkey, N. J., & Johnson, A. (2010). The effectiveness of an Acceptance and Commitment Therapy self-help intervention for chronic pain. *Clinical Journal of Pain*, 26, 393-402.

Juarascio, A. S., Forman, E. M., & Herbert, J. D. (2010). Acceptance and Commitment Therapy versus Cognitive Therapy for the treatment of co morbid eating pathology. *Behavior Modification*, 34, 175-190.

Smout, M., Longo, M., Harrison, S., Minniti, R., Wickes, W., & White, J. (2010). Psychosocial treatment for methamphetamine use disorders: a preliminary randomized controlled trial of cognitive behavior therapy and acceptance and commitment therapy. *Substance Abuse*, 31(2), 98-107.

Twohig, M. P., Hayes, S. C., Plumb, J. C., Pruitt, L. D., Collins, A. B., Hazlett-Stevens, H. & Woidneck, M. R. (2010) A randomized clinical trial of Acceptance and Commitment Therapy vs. Progressive Relaxation Training for obsessive compulsive disorder. Journal of Consulting and Clinical Psychology, 78, 705-716.

2011

Bohlmeijer, E. T., Fledderus, M., Rokx, T. A., & Pieterse, M. E. (2011). Efficacy of an early intervention based on acceptance and commitment therapy for adults with depressive symptomatology: Evaluation in a randomized controlled trial. Behaviour Research and Therapy, 49, 62-67.

Brinkborg, H., Michanek, J., Hesser, H., & Berglund, G. (2011). Acceptance and commitment therapy for the treatment of stress among social workers: A randomized controlled trial. Behaviour Research and Therapy, 49, 389-398.

Brown, L. A., Forman, E. M., Herbert, J. D., Hoffman, K. L., Yuen, E. K. and Goetter, E. M. (2011). A randomized controlled trial of acceptance-based behavior therapy and cognitive therapy for test anxiety: A pilot study. Behavior Modification, 35, 31-53.

Butryn, M. L., Forman, E., Hoffman, K., Shaw, J., & Juarascio, A. (2011). A pilot study of Acceptance and Commitment Therapy for promotion of physical activity. Journal of Physical Activity and Health, 8, 516-522.

Fledderus, M., Bohlmeijer, E.T., Pieterse, M. E., & Schreurs, K. M. (2011) Acceptance and commitment therapy as guided self-help for psychological distress and positive mental health: a randomized controlled trial. Psychological Medicine, 11, 1-11.

Gifford, E. V., Kohlenberg, B., Hayes, S. C., Pierson, H., Piasecki, M., Antonuccio, D., & Palm, K. (2011). Does acceptance and relationship focused behavior therapy contribute to bupropion outcomes? A randomized controlled trial of FAP and ACT for smoking cessation. Behavior Therapy, 42, 700-715.

Hayes, L., Boyd, C. P., & Sewell, J. (2011). Acceptance and Commitment Therapy for the treatment of adolescent depression: A pilot study in a psychiatric outpatient setting. Mindfulness, 2, 86-94.

Muto, T., Hayes, S. C., & Jeffcoat, T. (2011). The effectiveness of Acceptance and Commitment Therapy bibliotherapy for enhancing the psychological health of Japanese college students living abroad. Behavior Therapy, 42, 323–335.

Thorsell, J., Finnes, A., Dahl, J., Lundgren, T., Gybrant, M., Gordh, T., & Buhrman, M. (2011). A comparative study of 2 manual-based self-help interventions, Acceptance and Commitment Therapy and Applied Relaxation, for persons with chronic pain. The Clinical Journal of Pain, 49, 901-907.

Weineland, S., Arvidsson, D., Kakoulidis, T., & Dahl, J. (2011). Acceptance and commitment therapy for bariatric surgery patients, a pilot RCT. Obesity Surgery, 21, 1044-1044.

Westin, V. Z., Schulin, M., Hesser, H., Karlsson, M., Noe, R. Z., Olofsson, U., Stalby, M., Wisung, G. & Andersson, G. (2011). Acceptance and Commitment Therapy versus Tinnitus Retraining Therapy in the treatment of tinnitus distress: A randomized controlled trial. Behaviour Research and Therapy , 49, 737-747.

Wetherell JL, Afari N, Rutledge T, Sorrell JT, Stoddard JA, Petkus AJ, Solomon BC, Lehman DH, Liu L, Lang AJ, Hampton Atkinson J. (2011). A randomized, controlled trial of acceptance and commitment therapy and cognitive-behavioral therapy for chronic pain. Pain, 152, 2098-2107.

White, R.G., Gumley, A.I., McTaggart, J., Rattrie, L., McConville, D., Cleare, S, & Mitchell G. (2011). A feasibility study of Acceptance and Commitment Therapy for emotional dysfunction following psychosis. Behaviour Research and Therapy , 49, 901-907.

2012

Luoma, J. B., Kohlenberg, B. S., Hayes, S. C. & Fletcher, L. (2012). Slow and steady wins the race: A randomized clinical trial of Acceptance and Commitment Therapy targeting shame in substance use disorders. Journal of Consulting and Clinical Psychology, 80, 43-53.

Pearson, A. N., Follette, V. M. & Hayes, S. C. (2012). A pilot study of Acceptance and Commitment Therapy (ACT) as a workshop intervention for body dissatisfaction and disordered eating attitudes. Cognitive and Behavioral Practice, 19, 181-197.

In addition to outcome studies, at least 46 experimental psychopathology and analog component studies testing various aspects of the ACT model have been published in peer-reviewed journals. Typically, these studies compare the effects of one method of instantiating a single ACT process, such as defusion, acceptance, or (more broadly) mindfulness, on subject responses to aversive stimulation. For example, Masedo and Esteve (2007) investigated the effectiveness of acceptance versus suppression and spontaneous coping instructions during a cold pressor task (where subjects immerse their hands in ice cold water), and found that those in the acceptance condition reported increased pain tolerance and decreased pain ratings. While these 46 studies are of varying methodological quality, all lend support to various aspects of the ACT model. Additionally, at least 10 published, peer-reviewed ACT outcome studies have included formal meditational analyses uniformly concluding that ACT-specific processes were responsible for a significant amount of positive post-treatment client changes, with at least an additional 10 published studies using less formal meditational analyses reaching the same conclusion (see Baron & Kenny, 1986, for an independent discussion of meditational analysis). Though it is now somewhat dated, Hayes, Luoma, Bond, Masuda, and Lillis (2006) provided a review of ACT process and outcome evidence. Taken together with ACT's published outcome studies, the body of evidence supporting the ACT model and its general effectiveness is compelling, though additional research is needed.

The Links between ACT and RFT: Review of Published Studies

As of the time of this writing, studies that directly connect RFT phenomena with ACT processes are relatively limited. Factoring into this review is the ambiguity surrounding what kind of research qualifies as an RFT study (for review of one approach, see Dymond, May, Munnelly, & Hoon, 2010), much less an ACT-relevant RFT study. On one hand, a substantial volume of studies focusing on intervention components (Levin, Hildebrandt, Lillis, & Hayes, 2012) and rule-governance (Hayes, 1989) provide a general level of support for the RFT basis of ACT but typically lack an assessment of arbitrarily applicable derived relational responding. On the other hand, the volume of basic RFT studies involving arbitrarily applicable derived relational responding as the dependent variable is also substantial (for review, see Hayes, Barnes-Holmes, & Roche, 2001), but these studies often lack direct, clinically-relevant implications. For this review, we will focus on RFT studies where the assessment of derived relational responding (as may be done with the matching-to-sample procedure or related methods) relates to other measures of clinical relevance.

RFT studies on perspective taking and self touch upon the repertoires described by self-as-process and self-as-context. Compared to other core ACT processes, these may be some of the more robustly researched areas producing data with clinical implications, in part because self-concept is a known correlate of clinical symptoms. For example, Barnes, Lawlor, Smeets, and Roche (1996) demonstrated differential transformation of function in respect to evaluative, self-referring stimuli with intellectually disabled and non-disabled children. Similarly, Merwin and Wilson (2005) found that the ability to derive equivalence relations between self-referring and evaluative stimuli was related to self-reported levels of distress and self-esteem. Vahey, Barnes-Holmes, Barnes-Holmes, and Stewart (2009) showed that relational conditioning repertoires among a collection of self-referring and evaluative stimuli predicted belongingness to either a privileged or unprivileged block of prisoners. A study involving a sample of people exhibiting social anhedonia showed that performance on a deictic relational task was less accurate in comparison to normal controls (Villatte, Monestès, McHugh, Freixa i Baqué, & Loas, 2008), while a follow-up study replicated this effect among participants with high levels of social anhedonia as well as with participants diagnosed with schizophrenia (Villatte, Monestès, McHugh, Freixa i Baqué, & Loas, 2010).

RFT studies somewhat relevant to acceptance and defusion might focus on derived responding in respect to aversive stimulation as well as changes in those derived responses. A small collection of studies examining the issue of derived aversive control over behavior has been published. For example, arbitrarily applicable relational conditioning paradigms have been used to demonstrate that aversive stimulus functions may be derived via respondent and operant behavior through a variety of relational frames (Dougher, Hamilton, Fink, & Harrington, 2007; Dymond, Roche, Forsyth, Whelan, & Rhoden, 2008; Dymond et al., 2011; Rodríguez-Valverde, Luciano, & Barnes-Holmes, 2009; Rodríguez-Valverde, Luciano Soriano, Gutiérrez Martínez, & Hernández López, 2004). Hooper, Saunders, and McHugh (2010) have shown that a suppression strategy may generalize to stimuli bearing a derived avoidance function. Dougher, Augustson, Markham, Greenway, and Wulfert (1994) demonstrated derived acquisition as well as extinction of respondent elicitation. Roche, Kanter, Brown, Dymond, and Fogarty (2008) examined extinction of avoidance. Interestingly, this study indicated that extinction via presentation of a stimulus bearing derived aversive functions may be more effective than presentation of a stimulus bearing directly conditioned aversive functions, a finding that may have relevance to defusion interventions.

RFT studies relevant to values and committed action might investigate the role of derived motivational functions on behavior (also known as augmentals; Hayes, Zettle, & Rosenfarb, 1989). At least three studies have examined the derivation of consequential functions with RFT methods. Hayes, Kohlenberg, and Hayes (1991) demonstrated this transformation via relational frames of equivalence. Whelan and Barnes-Holmes (2004) demonstrated it through frames of same/opposite and Whelan, Barnes-Holmes, and Dymond (2006) demonstrated it via frames of more/less.

Finally, Hooper, Villatte, Neofotistou, and McHugh (2010) used the Implicit Relational Assessment Procedure (IRAP; Barnes-Holmes et al., 2006; Barnes-Holmes, Barnes-Holmes, Stewart, & Bowles, 2010) to assess the effects of mindfulness and thought suppression interventions on experiential avoidance, exemplifying a focus on multiple ACT processes (i.e., the mindfulness-based processes of defusion, acceptance, self-as-context, and contact with the present moment). Although the IRAP does not directly assess instances of derived relational responding, it was developed out of a body of RFT research and seems to detect particular biases in relational responding. Hooper, Villatte, et al. (2010) found the IRAP procedure to be more sensitive than self-report measures in detecting experiential avoidance, suggesting that the IRAP may hold promise for additional investigations of ACT-relevant processes.

Challenges in Linking RFT Studies to ACT's Six Core Processes

One key issue that requires redressing in order to maximize the interrelatedness of ACT and RFT research is a full reconciliation between mid-level ACT process terms and RFT-based processes. ACT's six core processes were designated largely for the benefit of clinicians (see, for example, Strosahl, Hayes, Wilson, & Gifford, 2004) to highlight, in a relatively accessible manner, what ACT clinicians should be doing on a functional level with clients. While preliminary theoretical accounts of the basic behavioral underpinnings of these processes have been offered (see, for example, Blackledge & Barnes-Holmes, 2009), it is arguably naïve to assume that the definitions of these midlevel processes will map seamlessly and sensibly onto basic RFT processes. Some midlevel terms (such as aspects of self-as-context) are readily operationalized in RFT terms, and indeed have already been studied in basic experimental labs (e.g., Dymond & Barnes, 1994, 1995, 1996, & 1997). Others, such as values and increased contact with the present moment, appear to be relatively removed from a basic RFT analysis. It seems plausible that the emergence of a full technical RFT-based account of active processes in ACT may involve such things as the operationalization of new technical terms that subsume two or more midlevel process terms, or perhaps even the introduction of a new technical process term that better enables ACT-relevant basic experimental research. We now move to an example of how a more RFT-friendly conceptualization of ACT might initially look.

A Process Description of Psychological Flexibility

Perhaps a generative strategy for increasing the empirical scrutiny of RFT-ACT links might begin with a more technical definition of psychological flexibility. In its current form, psychological flexibility is defined by a convergence of different albeit related behaviors. These behaviors are described in colloquial fashion, and their labels are regarded as midlevel terms. In what follows, we will suggest more technical ways to define these behaviors by referencing contextual conditions and resulting changes in behavior. We will conclude with a technical definition of psychological flexibility that summarizes these specifications.

Self-as-process. Present moment awareness, at a very basic level, must involve the simple operant behavior of discriminating particular features of the environment. The verbal nature of this discrimination adds a level of complexity, as the discrimination will involve the derivation and transformation of stimulus functions. In terms of perspective taking, the behavior involves a discrimination of events participating in relational frames of HERE/NOW from events participating in relational frames of THERE/THEN. If self-as-process is an active component of psychological flexibility, then performance on measures of flexibility would be facilitated by more accurate verbal discriminations between events HERE/NOW and events THERE/THEN. Alternatively, inaccurate performances will result in a greater degree of interference of stimulus functions coordinated with the frames of THERE/THEN, and an adverse performance on measures of flexibility.

Self-as-context. Observing the observer, similar to present moment awareness, involves a verbal discrimination of deictic frames (i.e., frames of perspective). Specifically, this behavior may be conceptualized as a consistent discrimination of the act of observing via relational frames of I/HERE/NOW about events THERE/THEN, regardless of content and especially in respect to events bearing significant approach or escape functions. All sources of stimulation are discriminated as occurring THERE/THEN except for the perspective from which observations are made. To the extent that self-as-context contributes to flexibility, performance on measures of flexibility would be facilitated by more prevalent repertoires of discriminating I/HERE/NOW about events THERE/THEN.

Defusion. Defusion may be conceptualized as the act of discriminating between verbal and nonverbal stimulus functions. Verbal reports that accurately tact the unconditioned stimulus functions of a verbal antecedent may qualify as defusion according to this definition. Better performance in distinguishing verbal and nonverbal stimulation may predict higher levels of flexibility. However, defusion may also be revealed by an effect on behavior—namely, a reduction in responding on the basis of verbal stimulus functions and/or an increase in responding on the basis of nonverbal stimulus functions. Such effects may be revealed through measures of respondent-like reactions (e.g., skin conductance, response latency, etc.) or assessment of operant responding.

Acceptance. Acceptance may be conceptualized as an approach response and/or the absence of an escape response in respect to aversive stimulation—unconditioned, conditioned, or derived. Existing studies have measured responding to graphic images, breath-holding, electric shock, exposure to ice water, and evaluative words. Choosing to remain in contact with such stimuli would be associated with higher degrees of psychological flexibility.

Values. Values have been discussed as augmentals, or verbal establishing operations (Dahl, Plumb, Stewart, & Lundgren, 2009). That is, values are verbal antecedents that function to increase the effectiveness of relevant consequences and the probability of any behavior that has generated those consequences, as well as the occasioning functions of antecedent stimuli that have been contingent on those consequences. According to Plumb, Stewart, Dahl, & Lundgren (2009), a value is a verbal behavior participating as the pinnacle event in a hierarchical network of verbal relations, such that a value encompasses many other aspects of life. In a sense, valuing may involve verbalizing, privately or publicly, a desired quality in activities across many areas of life. Given that motivating operations are defined in respect to contingent behavior and consequences, we will discuss methods of measuring values in respect to committed action.

Committed action. Perhaps of all the processes listed here, committed action seems to be the most overt example of a behavior. However, committed action is only meaningful in respect to values, or one's motivation for behaving. Thus, we will combine values and committed action into one coordinated definition. Valued action involves a verbal report (public or private) that functions as an augmental for a subsequent behavior that itself occasions a derived relation of coordination between the augmental and verbal discriminations of the behavior as it occurs. In other words, values-consistent action is verbally motivated behavior that is reinforced by, among other things, a sense of verbal coherence regarding the act and the motivation for committing that act.

Psychological flexibility. If we combine all six definitions, a more technical definition of psychological flexibility involves a variety of behaviors converging in one contingent stream of events. In accordance with these definitions, psychological flexibility involves:

1. a verbal report (public or private) that participates as the hierarch in a hierarchical derived relational network and functions as an augmental,

2. an approach response and/or absence of an escape, avoidance, or experiential avoidance response in respect to aversive or poten-

tially aversive stimulation that may be accompanied and facilitated by:

    a. discriminations between verbal and nonverbal stimulus functions,

    b. discriminations of events participating in relational frames of HERE/NOW from events in relational frames of THERE/THEN, and

    c. discriminations of the act of observing via relational frames of I/HERE/NOW about events THERE/THEN,

3. and a behavior that itself participates in a derived relation of coordination with the verbal report in criterion #1.

If we translate this definition back to the midlevel terminology of ACT's six core processes, psychological flexibility is an awareness of values and an act of acceptance that involves defusion, present moment awareness, and self-as-context to the extent that is necessary to result in committed action. We offer the more technical definition as a potential basis for a program of research, not as a definitive definition of psychological flexibility. This definition may arouse certain questions or objections. For example, would the sequence of these behaviors have any impact on overall flexibility? Must aversive stimulation be involved for behavior to be regarded as flexible? Do these divisions of behavior represent the optimal task analysis of psychological flexibility, or would further subdivisions and/or unifications be merited? We offer these questions not just as matters of conceptual debate, but also as some possible foci of research, so that the definition may be improved or discarded in the service of tightening the links between RFT and ACT.

ACT: Future Research Challenges

Within the past 10 years, the number of published randomized controlled trials of ACT has increased exponentially. However, the quality of many of these studies has been compromised (see, for example, Ost, 2008), at least in part due to often-severe funding limitations. This is quite common for an emerging treatment. High power, multisite outcome studies with a high degree of methodological rigor require large grants, and funding agencies are typically very reluctant to risk the millions of dollars required to fund such studies on treatments with relatively little empirical support. With ACT rapidly garnering increased empirical support, it now becomes a priority for such studies to be funded and conducted. As with any form of psychotherapy, it is also important that multiple research labs with divergent theoretical allegiances eventually conduct such studies to help minimize potential confounds such as experimenter bias.

Considerable work also needs to done with regard to pinning down ACT's conceptual and experimental roots. Though ACT and RFT emerged in tandem and a growing body of basic experimental evidence consistent with ACT theory is emerging, large aspects of ACT have not received the level of basic experimental scrutiny preferred for a behavioral theory. It would arguably be unfair to call this a deficiency relative to other models of psychotherapy given the status quo of clinical psychology, where the great majority of therapeutic modalities have little or no basic experimental evidence or even formal ties to basic experimental theory. But behaviorism has a strategic tradition of maintaining close ties between basic and applied research programs. Therapeutic techniques based on operant and classical conditioning, for example, emerged directly from the experimental laboratory, where hundreds (if not thousands) of tightly controlled experiments provided focused guidelines for changing human behavior. The rationale behind adopting such a model involves the assumption that doing so will give us a more timely and more specific understanding of what types of interventions impact behavior under various conditions. From the beginning, Hayes and colleagues sought to replicate this pattern with the ACT-RFT research program (see, for example, Levin & Hayes, 2009a), while acknowledging the additional difficulties that arise given the complexity of verbal behavior and the limitations basic researchers have on controlling the learning history of human subjects. As stated earlier in this chapter, this work has begun in earnest and is ongoing, but far from complete.

References

Bach, P. A., & Moran, D. J. (2008). ACT in Practice. Oakland, CA: New Harbinger.

Barnes, D., Lawlor, H., Smeets, P. M., & Roche, B. (1996). Stimulus equivalence and academic self-concept among mildly mentally handicapped and nonhandicapped children. *Psychological Record, 46,* 87-107.

Barnes-Holmes, D., Barnes-Holmes, Y., Power, P., Hayden, E., Milne, R., & Stewart, I. (2006). Do you really know what you believe? Developing the Implicit Relational Assessment Procedure (IRAP) as a direct measure of implicit beliefs. *Irish Psychologist, 32,* 169-177.

Barnes-Holmes, D., Barnes-Holmes, Y., Stewart, I., & Bowles, S. (2010). A sketch of the Implicit Relational Assessment Procedure (IRAP) and the Relational Elaboration and Coherence (REC) model. *Psychological Record, 60,* 527-542.

Barnes-Holmes, D., Hayes, S. C., & Dymond, S. (2001). Self and self-directed rules. In S. Hayes, D. Barnes-Holmes, & B. Roche (Eds.), *Relational frame theory: A post-Skinnerian account of human language and cognition* (pp. 119-140). New York: Kluwer Academic.

Baron, R. M., & Kenny, D. A. (1986). The moderator-mediator variable distinction in social psychological research: Conceptual, strategic, and statistical considerations. *Journal of Personality and Social Psychology, 51,* 1173–1182.

Beck, A. T. (1967). *Depression: Clinical, experimental, and theoretical aspects.* New York: Hoeber. Republished as *Depression: Causes and treatment.* Philadelphia: University of Pennsylvania Press).

Beck, A. Rush, A., Shaw, B., & Emery, G. (1979). *The cognitive therapy of depression.* New York: Guilford Press.

Blackledge, J. T. (2007). Disrupting verbal processes: Cognitive defusion in Acceptance and Commitment Therapy and other mindfulness-based psychotherapies. *The Psychological Record, 57(4),* 555-576.

Blackledge, J. T., & Barnes-Holmes, D. (2009). Core processes in acceptance and commitment therapy. In J. T. Blackledge, J. Ciarrochi, & F. P. Deane (Eds.), *Acceptance and Commitment Therapy: Contemporary theory research and practice* (pp. 1-40). Sydney: Australian Academic Press.

Brinkborg, H., Michanek, J., Hesser, H., & Berglund, G. (2011). Acceptance and commitment therapy for the treatment of stress among social workers: A randomized controlled trial. *Behaviour Research and Therapy, 49,* 389-398.

Chomsky, N. (1957). A review of B. F. Skinner's *Verbal Behavior. Language, 35 (1),* 26-58.

Clark, A., & Chalmers, D. J. (1998). The extended mind. *Analysis 58,* 10-23.

Dahl, J. C., Plumb, J. C., Stewart, I., & Lundgren, T. (2009). *The art and science of valuing in psychotherapy.* Oakland, CA: New Harbinger.

Dahl, J., Wilson, K. G., & Nilsson, A. (2004). Acceptance and Commitment Therapy and the treatment of persons at risk for long-term disability resulting from stress and pain symptoms: A preliminary randomized trial. *Behavior Therapy, 35,* 785-802.

Dahl, J., Wilson, K. G., & Nilsson, A. (2004). Acceptance and Commitment Therapy and the treatment of persons at risk for long-term disability resulting from stress and pain symptoms: A preliminary randomized trial. *Behavior Therapy, 35*, 785-802.

Dougher, M. J., Augustson, E., Markham, M. R., Greenway, D. E., & Wulfert, E. (1994). The transfer of respondent eliciting and extinction functions through stimulus equivalence classes. *Journal of the Experimental Analysis of Behavior, 62*, 331-351.

Dougher, M. J., Hamilton, D. A., Fink, B. C., & Harrington, J. (2007). Transformation of the discriminative and eliciting functions of generalized relational stimuli. *Journal of the Experimental Analysis of Behavior, 88*, 179-197.

Dowd, E. T. (2004). Cognition and the cognitive revolution in psychotherapy: Promises and advances. *Journal of Clinical Psychology, 60(4)*, 415-428.

Dryden, W., & Ellis, A. (1986). Rational-emotive therapy (RET). In W. Dryden & W. Golden (Eds.), *Cognitive-behavioural approaches to psychotherapy* (pp. 129-168). London: Harper & Row.

Dymond, S., & Barnes, D. (1994). A transfer of self-discrimination response functions through equivalence relations. *Journal of the Experimental Analysis of Behavior, 62(2)*, 251-267.

Dymond, S., & Barnes, D. (1995). A transformation of self-discrimination response functions in accordance with the arbitrarily applicable relations of sameness, more-than, and less-than. *Journal of the Experimental Analysis of Behavior, 64*, 163-184.

Dymond, S., & Barnes, D. (1996). A transformation of self-discrimination response functions in accordance with the arbitrarily applicable relations of sameness and opposition. *Psychological Record, 46(2)*, 271-300.

Dymond, S., & Barnes, D. (1997). Behavior analytic approaches to self-awareness. *Psychological Record, 47*, 181-200.

Dymond, S., May, R. J., Munnelly, A., & Hoon, A. E. (2010). Evaluating the evidence base for Relational Frame Theory: A citation analysis. *Behavior Analyst, 33*, 97-117.

Dymond, S., Roche, B., Forsyth, J. P., Whelan, R., & Rhoden, J. (2008). Derived avoidance learning: Transformation of avoidance response functions in accordance with same and opposite relational frames. *Psychological Record, 58*, 269-286.

Dymond, S., Schlund, M. W., Roche, B., Whelan, R., Richards, J., & Davies, C. (2011). Inferred threat and safety: Symbolic generalization of human avoidance learning. *Behaviour Research and Therapy, 49*, 614-621.

Flaxman, P. E., & Bond, F. W. (2010). A randomized worksite comparison of acceptance and commitment therapy and stress inoculation training. *Behaviour Research and Therapy 43*, 816-820.

Forman, E. M., Herbert, J. D., Moitra, E., Yeomans, P. D. & Geller, P. A. (2007). A randomized controlled effectiveness trial of Acceptance and Commitment Therapy and Cognitive Therapy for anxiety and depression. *Behavior Modification, 31(6)*, 772-799.

Gaudiano, B. A. (2009). Öst's (2008) methodological comparison of clinical trials of acceptance and commitment therapy versus cognitive behavior therapy: Matching apples with oranges? *Behaviour Research and Therapy, 47,* 1066-1070.

Gaudiano, B. A. (2011). A review of acceptance and commitment therapy (ACT) and recommendations for continued scientific advancement. *The Scientific Review of Mental Health Practice, 8,* 5-22.

Gifford, E. V., & Hayes, S. C. (1999). Functional contextualism: A pragmatic philosophy for behavioral science. In W. O'Donohue & R. Kitchener (Eds.), *Handbook of behaviorism* (pp. 285-327). San Diego: Academic Press.

Gifford, E. V., Kohlenberg, B. S., Hayes, S. C., Antonuccio, D. O., Piasecki, M. M., Rasmussen-Hall, M. L., & Palm, K. M. (2004). Acceptance based treatment for smoking cessation: An initial trial of Acceptance and Commitment Therapy. *Behavior Therapy, 35,* 689-705.

Gifford, E. V., Kohlenberg, B., Hayes, S. C., Pierson, H., Piasecki, M., Antonuccio, D., & Palm, K. (2011). Does acceptance and relationship focused behavior therapy contribute to bupropion outcomes? A randomized controlled trial of FAP and ACT for smoking cessation. *Behavior Therapy, 42,* 700-715.

Gratz, K. L., & Gunderson, J. G. (2006). Preliminary data on an acceptance-based emotion regulation group intervention for deliberate self-harm among women with Borderline Personality Disorder. *Behavior Therapy, 37,* 25-35.

Gregg, J. A., Callaghan, G. M., Hayes, S. C., & Glenn-Lawson, J. L. (2007). Improving diabetes self-management through acceptance, mindfulness, and values: A randomized controlled trial. *Journal of Consulting and Clinical Psychology, 75*(2), 336-343.

Hatfield, G. (2002). Psychology, philosophy, and cognitive science: Reflections on the history and philosophy of experimental psychology. *Mind and Language, 17*(3), 207-232.

Hayes, S. C. (1987). A contextual approach to therapeutic change. In N. Jacobsen (Ed.), *Psychotherapists in clinical practice: Cognitive and behavioral perspectives* (pp. 327-387). New York: Guilford.

Hayes, S. C. (1989). *Rule-governed behavior: Cognition, contingencies, and instructional control.* New York: Plenum.

Hayes, S. C., Strosahl, K., & Wilson, K. G. (2011). *Acceptance and Commitment Therapy: The process and practice of mindful change* (2nd edition). New York: Guilford Press.

Hayes, S. C., Bissett, R., Roget, N., Padilla, M., Kohlenberg, B. S., Fisher, G., Masuda, A., Pistorello, J., Rye, A. K., Berry, K. & Niccolls, R. (2004). The impact of acceptance and commitment training and multicultural training on the stigmatizing attitudes and professional burnout of substance abuse counselors. *Behavior Therapy, 35,* 821-835.

Hayes, S. C., Kohlenberg, B. K., & Hayes, L. J. (1991). The transfer of specific and general consequential functions through simple and conditional equivalence classes. *Journal of the Experimental Analysis of Behavior, 56,* 119-137.

Hayes, S. C., Kohlenberg, B. S., & Melancon, S. M. (1989). Avoiding and altering rule-control as a strategy of clinical intervention. In S. C. Hayes (Ed.), *Rule-governed behavior: Cognition, contingencies, and instructional control* (pp. 359-385). New York: Plenum.

Hayes, S. C., Luoma, J., Bond, F., Masuda, A., & Lillis, J. (2006). Acceptance and Commitment Therapy: Model, processes, and outcomes. *Behaviour Research and Therapy, 44*(1), 1-25.

Hayes, S. C., & Strosahl, K. (Eds.). (2004). *A practical guide to acceptance and commitment therapy.* New York: Springer.

Hayes, S. C., Strosahl, K. D. Bunting, K., Twohig, M. P., & Wilson, K. G. (2004). What is acceptance and commitment therapy? In S. C. Hayes & K. D. Strosahl (Eds.), *A Practical Guide to Acceptance and Commitment Therapy* (pp. 1-30). New York: Guilford Press.

Hayes, S. C., Strosahl, K., & Wilson, K. G. (1999). *Acceptance and commitment therapy: An experiential approach to behavior change.* New York: Guilford.

Hayes, S. C., Barnes-Holmes, D., & Roche, B. (2001). *Relational Frame Theory: A Post-Skinnerian account of human language and cognition.* New York: Plenum Press.

Hayes, S. C., Villatte, M., Levin, M. & Hildebrandt, M. (2011). Open, aware, and active: Contextual approaches as an emerging trend in the behavioral and cognitive therapies. *Annual Review of Clinical Psychology, 7,* 141-168.

Hayes, S. C., Wilson, K. G., Gifford, E. V., Bissett, R., Piasecki, M., Batten, S. V., Byrd, M., & Gregg, J. (2004). A randomized controlled trial of twelve-step facilitation and acceptance and commitment therapy with polysubstance abusing methadone maintained opiate addicts. *Behavior Therapy, 35,* 667-688.

Hayes, S. C., Wilson, K. W., Gifford, E. V., Follette, V. M., & Strosahl, K. (1996). Experiential avoidance and behavioral disorders: A functional dimensional approach to diagnosis and treatment. *Journal of Consulting and Clinical Psychology, 64,* 1152-1168.

Hayes, S. C., Zettle, R. D., & Rosenfarb, I. (1989). Rule following. In S. C. Hayes (Ed.), *Rule-governed behavior: Cognition, contingencies, and instructional control* (pp. 191-220). New York: Plenum.

Hollon, S. (2010). Aaron T. Beck: The cognitive revolution in theory and therapy. In Castonguay, C. Muran, & L. Angus (Eds.), *Bringing psychotherapy research to life: Understanding change through the work of leading clinical researchers.* Washington DC: American Psychological Association.

Hollon, S. D., & Beck, A. T. (1979). Cognitive therapy of depression. In P. C. Kendall & S. D. Barlow (Eds.), *Cognitive-behavioral Intervention: Theory, Research, and Procedures* (pp.153-203). New York: Academic Press.

Hooper, N., Saunders, S., & McHugh, L. (2010). The derived generalization of thought suppression. *Learning and Behavior, 38,* 160-168.

Hooper, N., Villatte, M., Neofotistou, E., & McHugh, L. (2010). The effects of mindfulness versus thought suppression on implicit and explicit measures of experimental avoidance. *International Journal of Behavior Consultation and Therapy, 6,* 233-244.

Johnston, M., Foster, M., Shennan, J., Starkey, N. J., & Johnson, A. (2010). The effectiveness of an Acceptance and Commitment Therapy self-help intervention for chronic pain. *Clinical Journal of Pain, 26,* 393-402.

Levin, M. E., & Hayes, S. C. (2009a). ACT, RFT, and contextual behavioral science. In J. T. Blackledge, J. Ciarrochi, & F. P. Deane (Eds.), *Acceptance and Commitment Therapy: Contemporary research and practice* (pp. 1-40). Sydney: Australian Academic Press.

Levin, M., & Hayes, S. C. (2009b). Is acceptance and commitment therapy superior to established treatment comparisons? *Psychotherapy and Psychosomatics, 78,* 380.

Levin, M. E., Hildebrandt, M. J., Lillis, J., & Hayes, S. C. (2012). The impact of treatment components suggested by the psychological flexibility model: A meta-analysis of laboratory-based component studies. *Behavior Therapy, 43,* 741-756.

Lillis, J., Hayes, S. C., Bunting, K., & Masuda, A. (2009). Teaching acceptance and mindfulness to improve the lives of the obese: A preliminary test of a theoretical model. *Annals of Behavioral Medicine, 37,* 58-69.

Lundgren, A. T., Dahl, J., Melin, L. & Kees, B. (2006). Evaluation of Acceptance and Commitment Therapy for drug refractory epilepsy: A randomized controlled trial in South Africa. *Epilepsia, 47,* 2173-2179.

Lundgren, T., Dahl, J., Yardi, N., & Melin, L. (2008). Acceptance and Commitment Therapy and yoga for drug-refractory epilepsy: A randomized controlled trial. *Epilepsy & Behavior, 13,* 102–108.

Masedo, A. I., & Esteve, M. R. (2007). Effects of suppression, acceptance and spontaneous coping on pain tolerance, pain intensity and distress. *Behaviour Research and Therapy, 45,* 199-209.

Merwin, R. M., & Wilson, K. G. (2005). Preliminary findings on the effects of self-referring and evaluative stimuli on stimulus equivalence class formation. *Psychological Record, 55,* 561-575.

Öst, L. (2008). Efficacy of the third wave of behavioral therapies: A systematic review and meta-analysis. *Behaviour Research and Therapy, 46*(3), 296-321.

Páez, M. B., Luciano, C., & Gutiérrez, O. (2007).Tratamiento psicológico para el afrontamiento del cáncer de mama. Estudio comparativo entre estrategias de aceptación y de control cognitivo. *Psicooncología, 4,* 75-95.

Pepper, S. (1961). *World hypotheses: A study in evidence.* Berkeley, CA: University of California Press.

Petersen, C. L., & Zettle, R. D. (2009). Treating inpatients with comorbid depression and alcohol use disorders: A comparison of Acceptance and Commitment Therapy and treatment as usual. *The Psychological Record, 59,* 521-536.

Plumb, J. C., Stewart, I., Dahl, J., & Lundgren, T. (2009) In search of meaning: Values in modern clinical behavior analysis. *Behavior Analyst, 32,* 85-103.

Roche, B. T., Kanter, J. W., Brown, K. R., Dymond, S., & Fogarty, C. C. (2008). A comparison of "direct" versus "derived" extinction of avoidance responding. *Psychological Record, 58,* 443-464.

Rodríguez-Valverde, M., Luciano, C., & Barnes-Holmes, D. (2009). Transfer of aversive respondent elicitation in accordance with equivalence relations. *Journal of the Experimental Analysis of Behavior, 92,* 85-111.

Rodríguez-Valverde, M., Luciano Soriano, M. C., Gutiérrez Martínez, O., & Hernández López, M. (2004). Transfer of latent inhibition of aversively conditioned autonomic responses through equivalence classes. *International Journal of Psychology and Psychological Therapy, 4,* 605-622.

Skinner, B. F. (1953). *Science and Human Behavior.* New York: Macmillan.

Skinner, B. F. (1969). *Contingencies of reinforcement: A theoretical analysis.* New York: Appleton.

Skinner, B. F. (1992). *Verbal behavior.* Acton, Massachusetts: Copley Publishing Group.

Stepp, N., Chemero, A., & Turvey, M. T. (2011). Philosophy for the rest of cognitive science. *Topics in Cognitive Science, 3,* 425-437.

Stewart, I., & Barnes-Holmes, D. (2004). Relational frame theory and analogical reasoning: Empirical investigations. *International Journal of Psychology and Psychological Therapy, 4,* 241-262.

Strosahl, K., Hayes, S. C., Wilson, K. G., & Gifford, E. V. (2004). An ACT primer: Core Therapy processes, intervention strategies, and therapist competencies. In S. Hayes & K. Strosahl (Eds.), *A practical guide to acceptance and commitment therapy* (pp. 31-58). New York: Springer.

Tapper, K., Shaw, C., Ilsley, J., Hill, A. J., Bond, F. W., & Moore, L. (2009). Exploratory randomized controlled trial of a mindfulness-based weight loss intervention for women. *Appetite, 52,* 396–404.

Tataryn, D. J., Nadel, L., & Jacobs, W. J. (1989). Cognitive therapy and cognitive science. In A. Freeman, K. Simon, L. Beutler, & H. Arkowitz (Eds.), *Comprehensive handbook of cognitive therapy* (pp. 83-98). New York: Plenum Press.

Thorsell, J., Finnes, A., Dahl, J., Lundgren, T., Gybrant, M., Gordh, T., & Buhrman, M. (2011). A comparative study of 2 manual-based self-help interventions, Acceptance and Commitment Therapy and Applied Relaxation, for persons with chronic pain. *The Clinical Journal of Pain, 27,* 716-723.

Vahey, N. A., Barnes-Holmes, D., Barnes-Holmes, Y., & Stewart, I. (2009). A first test of the Implicit Relational Assessment Procedure (IRAP) as a measure of self-esteem: Irish prisoner groups and university students. *The Psychological Record, 59,* 371–388.

Varra, A. A., Hayes, S. C., Roget, N., & Fisher, G. (2008). A randomized control trial examining the effect of Acceptance and Commitment Training on clinician willingness to use evidence-based pharmacotherapy. *Journal of Consulting and Clinical Psychology, 76,* 449-458.

Villatte, M., Monestès, J. L., McHugh, L., Freixa i Baqué, E., & Loas, G. (2008). Assessing deictic relational responding in social anhedonia: A functional approach to the development of Theory of Mind impairments. *International Journal of Behavioral Consultation and Therapy, 4,* 360-373.

Villatte, M., Monestès, J. L., McHugh, L., Freixa i Baqué, E., & Loas, G. (2010). Adopting the perspective of another in belief attribution: Contribution of Relational Frame Theory to the understanding of impairments in schizophrenia. *Journal of Behavior Therapy and Experimental Psychiatry, 41,* 125-134.

Vowles, K. E., McNeil, D. W., Gross, R. T. McDaniel, M. L., Mouse, A., Bates, M., Gallimore, P., & McCall, C. (2007). Effects of pain acceptance and pain control strategies on physical impairment in individuals with chronic low back pain. *Behavior Therapy, 38,* 412-425.

Westin, V. Z., Schulin, M., Hesser, H., Karlsson, M., Noe, R. Z., Olofsson, U., Stalby, M., Wisung, G. & Andersson, G. (2011). Acceptance and Commitment Therapy versus Tinnitus Retraining Therapy in the treatment of tinnitus: A randomized controlled trial. *Behaviour Research and Therapy, 49,* 737-747.

Wetherell, J. L., Afari, N., Rutledge, T., Sorrell, J. T., Stoddard, J. A., Petkus, A. J., Solomon, B. C.,Lehman, D. H., Liu, L., Lang, A. J., Hampton-Atkinson, J. (2011). A randomized, controlled trial of acceptance and commitment therapy and cognitive-behavioral therapy for chronic pain. *Pain, 152,* 2098-2107.

Whelan, R., & Barnes-Holmes, D. (2004). Empirical models of formative augmenting in accordance with the relations of same, opposite, more-than, and less-than. *International Journal of Psychology and Psychological Therapy, 4,* 285-302.

Whelan, R., Barnes-Holmes, D., & Dymond, S. (2006). The transformation of consequential functions in accordance with the relational frames of more-than and less-than. *Journal of the Experimental Analysis of Behavior, 86,* 317-335.

Wicksell, R, K., Ahlqvist, J., Bring, A., Melin, L. & Olsson, G. L. (2008). Can exposure and acceptance strategies improve functioning and quality of life in people with chronic pain and whiplash associated disorders (WAD)? A randomized controlled trial. *Cognitive Behaviour Therapy, 37,* 1-14.

Woods, D. W., Wetterneck, C. T., & Flessner, C. A. (2006) A controlled evaluation of Acceptance and Commitment Therapy plus habit reversal for trichotillomania. *Behaviour Research and Therapy, 44,* 639-656.

Zettle, R. D. (2003). Acceptance and commitment therapy (ACT) versus systematic desensitization in treatment of mathematics anxiety. *The Psychological Record, 53,* 197-215.

Zettle, R. (2005). The evolution of a contextual approach to therapy: From comprehensive distancing to ACT. *International Journal of Behavioral and Consultation Therapy, 1(2),* 77-89.

Zettle, R. D., & Hayes, S. C. (1982). Rule governed behavior: A potential theoretical framework for cognitive behavior therapy. In P. C. Kendall (Ed.), *Advances in cognitive behavioral research and therapy* (pp. 73-118). New York: Academic.

Zettle, R. D. & Hayes, S. C. (1986). Dysfunctional control by client verbal behavior: The context of reason giving. *The Analysis of Verbal Behavior, 4,* 30-38.

Zettle, R. D., & Rains, J. C. (1989). Group cognitive and contextual therapies in treatment of depression. *Journal of Clinical Psychology, 45,* 436-445.

CHAPTER ELEVEN

Putting Relational Frame Theory (RFT) to Work: Current and Future RFT Research in Organizational Behavior Management

Denis O'Hora

National University of Ireland, Galway

Kristen Maglieri

Trinity College Dublin

Triona Tammemagi

National University of Ireland, Galway

The workplace is a complex and ever-changing environment. The defining characteristic of all organizations is that individuals work together to achieve a common goal. Organizations depend on interconnected behavior by various people, because work is not produced in isolation (e.g., people work together to produce a product or service and deliver that to its customers). This interdependence is at the heart of the complexity of the influences on human behavior in organizations. Each organization has a unique culture, a set of values and practices that distinguish it from other organizations. These values and practices (e.g., recruitment, incentive schemes, performance appraisal systems, labor relations) contribute to the personal environment of each employee at each level of the organization. The contingencies within this personal environment determine whether an employee is creative, engaged, and productive, or frustrated, cynical and uninterested.

Organizational behavior management (OBM), the application of behavioral principles to organizational behavior (Bucklin, Alvero, Dickinson & Jackson, 2000), attempts to identify the critical contingencies in the employee's environment that can be adjusted to enhance employee satisfaction and productivity. The promise of OBM is that, by providing functional analyses of ineffective organizational contingencies, interventions can be designed to transform those contingencies into effective systems. To date, most research in OBM has been influenced by accounts of organizational contingencies that focus on direct reinforcement and punishment. However, for many years now, OBM researchers have recognized that much employee behavior is indirectly controlled by verbal statements of one kind or another (e.g., rules). People at work, both managers and employees, constantly exhibit complex verbal behaviors: thoughts, feelings, beliefs, attitudes, judgments, biases and values. Workers' behavior, verbal or non-verbal, covert or overt, is constantly changing and influenced by multiple sources. Organizational contingencies are thus exceedingly intricate.

Relational frame theory (RFT; Hayes, Barnes-Holmes & Roche, 2001) is particularly suited, among behavioral theories of language, to address the complexities of the workplace environment and behavior. This is because relational frame theory explicitly provides a behavioral account of reference; that is, how words "mean" what they "stand for". In this way, RFT articulates how such "meanings" can establish novel behaviors and novel consequences for individuals in the work environment. Specifically, RFT proposes that, through transformation of function, understanding verbal statements (e.g., rules) changes how stimuli in our environment affect us, which changes our behavior (for more detail, see Stewart & Roche, this volume). For instance, understanding that "quarterly returns" are important (i.e., they predict positive and/or negative reinforcement) may encourage employees to engage in behaviors that maximize such returns and to neglect other behaviors.

In this chapter, we review recent research on organizational behavior influenced by RFT. This research can be broadly construed within three themes: the application of Acceptance and Commitment Therapy (ACT) in organizational settings, the impact of psychological flexibility on employee health and performance, and the analysis of the effects of common organizational interventions. The practical application of RFT in the workplace is most readily seen in the application of therapeutic interventions based on ACT (Hayes, 1987; see also Blackledge & Drake, this volume). Working in an organization exposes individuals to overt physical dangers, such as the probability of injury on construction sites, but also less obvious psychological hazards such as stress and burnout. Interventions based on ACT have proven to be effective in the amelioration of psychological distress (e.g., depression, anxiety; Hayes, Masuda, Bissett, Luoma, & Guerrero, 2004). The first section of this review outlines a number of studies that have demonstrated that ACT interventions can be employed to reduce the severity and prevalence of psychological distress and enhance worker performance.

ACT Interventions in the Workplace

Bond and Bunce (2000) reported the first evaluation of an ACT intervention in the workplace (i.e., a large media organization). The study included a total of 90 participants in three groups. One group received an ACT intervention, the second received an Innovation Promotion Program (IPP), and the third served as a wait-list control group. The interventions were delivered across 3 months and lasted 9 hours. Both the ACT intervention and the IPP intervention significantly improved general mental health, depression, and the workers' propensity to innovate compared to the control group. Since Bond and Bunce's study, there have been numerous implementations of ACT interventions in a variety of organizations. For example, Dahl, Wilson and Nilsson (2004) conducted an evaluation of an ACT intervention on employees who were at risk to be put on long-term disability due to stress and pain symptoms. Using a randomized control trial, 19 public health workers were randomly assigned to either an ACT intervention group (with treatment as usual) or a treatment as usual group only (i.e., use of pain medication). The brief four-hour ACT intervention led to a reduction in the number of sick days and fewer medical visits at post-treatment and at 6 month follow-up compared to the control condition. More recently, Flaxman and Bond (2010) compared ACT with Stress Inoculation Training (SIT; a traditional CBT model) to reduce psychological distress for 107 employees of two large government organizations with above average levels of stress. Employees were randomly assigned to either an ACT, SIT or control group. Both the ACT and the SIT interventions were successful in improving general mental health compared to the control group.

Acceptance and Commitment Therapy interventions have been particularly successful in reducing worker burnout. Burnout is a combination of exhaustion, cynicism/depersonalization, and lack of personal accomplishment that comes about as the result of prolonged exposure to job stressors (Maslach, Schaufeli, & Leiter, 2001). Hayes, Bissett and colleagues (2004) compared the effects of an ACT intervention and multicultural training on burnout and stigmatizing attitudes of substance abuse counselors toward their clients. The interventions consisted of day-long workshops. Participants were assessed at the beginning of the workshop, at the end of the workshop, and at 3-month follow-up. Stigmatizing attitudes in the ACT condition did not improve at post-treatment, but did significantly improve by 3-month follow-up. Interestingly, the opposite occurred with the multicultural training group. Their stigmatizing attitudes improved at post-treatment, but not at follow-up. There was a significant improvement in burnout within the ACT group at post-treatment and follow-up.

A number of further studies have reported that ACT interventions can reduce burnout. Ruiz, Rios, and Martin (2008) described an evaluation of an ACT-based intervention in a Spanish hospital. The intervention was found to be effective in preventing the development of burnout among workers in a palliative care unit. The effect was particularly evident in terms of reducing feelings of depersonalization and increasing feelings of personal accomplishment. Brinkborg, Michanek, Hesser and Berglund (2011) investigated the effect of a brief ACT training as an intervention with 106 social workers experiencing stress. One third of the sample had been on sick leave for an extended time in the past, and in more than half of the cases this was for stress-related reasons. At post-treatment, the ACT group had significantly lower levels of perceived stress, and significantly fewer mental health problems than the control group. The ACT group had significantly fewer burnout symptoms, but there were no significant differences found in performance-based self-esteem, or job demand and control. Brinkborg et al. then separated the participants' data into high and low stress level groups based on their baseline stress level, and found that participants with different levels of stress responded differently to the ACT intervention. This analysis found that, in the high stress group, a substantial proportion (42%) improved to a clinically significant degree, but those with low pre-intervention levels of stress did not significantly improve. Finally, in a recent study by Lloyd, Bond and Flaxman (in press), 100 government customer service workers were randomly assigned to an ACT group or a wait-list control group. The ACT group received three 3-hour training sessions, two on consecutive weeks and one 2 months later. Measures were taken at the beginning of the first workshop (T1), the beginning of the second workshop (T2), the beginning of the third workshop (T3), and again 6 months after the final training workshop (T4). Results showed that there were significant reductions in burnout and strain in the ACT group compared to the control group. These results were apparent at T3, and maintained at T4 for depersonalization and emotional exhaustion.

ACT interventions have also been employed to enhance employee performance. Luoma et al. (2007) investigated whether adding an ACT group consultation to a standard 1-day training course would increase adoption of a Group Drug Counseling (GDC) technology, a group-therapy approach with empirical support. All 30 participants attended a 1-day 6-hr training workshop on how to implement GDC. Following this, 16 of these were assigned to an ACT group and participated in eight weekly 1.5hr training groups based on ACT and a Relapse Prevention Model (Marlatt & Gordon, 1985), focusing on overcoming barriers to implementing the GDC. The group exposed to the eight-week course was significantly more likely to adopt these new practices. Varra, Hayes, Roget, and Fisher (2008) evaluated the effectiveness of ACT training on 59 drug and alcohol counselors' use of evidence-based pharmacotherapy. Participants in the ACT group attended a 6-hour ACT training day, and the control group attended a 6-hour educational control training, before both groups attended the same 2-day training on evidence-based pharmacotherapy treatments. Participants with ACT pre-training showed significantly higher rates of referrals to pharmacotherapy at post-training and follow-up than those with educational control pre-training.

The Role of Psychological Flexibility

The foregoing research provides ample evidence that therapeutic interventions based on ACT are effective in both increasing worker resilience and enhancing innovation and performance. In order to understand the role that relational framing plays in these interventions, it is necessary to consider *how* these interventions work. ACT interventions are designed to impact an individual's psychological flexibility. Psychological flexibility is the ability to contact the present moment without avoidance, enabling persistence or change in behavior in pursuit of values or goals (Hayes, Luoma, Bond, Lillis, & Masuda, 2006). In ACT interventions, psychological flexibility is established through six core processes; acceptance, cognitive defusion, being in the present moment, self-as-context, values and committed action (Blackledge & Drake, current volume). Though a detailed consideration of the core processes of ACT is beyond the scope of the current chapter, readers who wish to learn more about the therapeutic processes of ACT and the practicalities of developing ACT-based interventions for the workplace are directed to Bond, Flaxman, van Veldhoven, and Biron (2010). Bond et al. provide a thorough introduction to these processes and comprehensive detail on the implementation of ACT-based interventions in organizations, including examples of dialogues between trainers and employees.

According to RFT, psychological flexibility can be viewed as an indication of an individual's repertoire of relational responding. Specifically, certain repertoires of relational responding enable more flexible, adaptive responding to environmental contingencies, whereas other repertoires result in less flexible responding (e.g., rigid rule following). From an RFT perspective, many of the functions of antecedents and consequences in the workplace are established through relational responding. For instance, the positive emotional feelings (or predicted feelings) associated with having a "corner office" might be due to its participation in sameness (coordination) relations with being successful ("I've made it"), or due to "greater than" (comparison) relations between the employee's view of himself/herself and his/her colleagues ("I'm the best", "I'm top dog"). Without language to make one office "better than" another, it is unlikely that the actual differences between offices would affect behavior. Rather, it is the derived sameness relations between the "corner office" and "success" established by the employee's verbal behavior and derived comparison relations between "me" and "them" that makes the "corner office" a derived reinforcer. Such verbal transformations of function are necessary for us to work together toward long-term goals. However, this power of language to transform the functions of stimuli may also give rise to psychological inflexibility because, to a degree, verbal behavior can establish its own reinforcement. In fact, numerous basic research studies have shown that verbal rules can induce behavior that is insensitive to programmed contingencies (see Hayes, Zettle & Rosenfarb, 1989 for a review).

A number of the intervention studies mentioned previously provided evidence that ACT interventions work by enhancing psychological flexibility. Bond and Bunce (2000) included the *Acceptance and Action Questionnaire* (AAQ; Hayes, Strosahl, et al., 2004) as a measure of psychological flexibility and found that scores on the AAQ mediated positive effects of their ACT intervention. Flaxman and Bond (2010) also measured psychological flexibility and found that, in the ACT group, change in psychological flexibility mediated the effect of the ACT intervention on general mental health when controlling for change in dysfunctional cognitions. Increased psychological flexibility thus functioned as a mediator of change even after controlling for changes in cognitive content. In the study by Lloyd et al. (in press) on burnout, increases in psychological flexibility in the ACT group mediated decreases in emotional exhaustion. These decreases in emotional exhaustion mediated the maintenance of depersonalization levels. This pattern of results suggests that the impact of psychological flexibility is not always direct, because, in this study, psychological flexibility beneficially impacted depersonalization through its effect on emotional exhaustion.

There is now considerable evidence that psychological flexibility is an important predictor of employee mental health. Bond and Bunce (2003) conducted a longitudinal study with 412 customer service center workers. The AAQ was used as a measure of psychological flexibility, and the authors also measured job control using a Job Control Questionnaire (Ganster, 1989). Outcome variables included computer input errors, scores on the GHQ and scores on a general job satisfaction scale. Greater flexibility not only predicted better mental health (GHQ), but also performance (lower errors) one year later. Donaldson-Feilder and Bond (2004) surveyed 290 workers in the UK to compare how well psychological flexibility and emotional intelligence, another meta-cognitive measure, predicted well-being. These employees came from five different organizations: a manufacturing company based on the south coast of England; the London office of an overseas government; the management consultancy arm of a large accountancy firm; the corporate headquarters of an insurance broker; and a financial services consultancy. The correlation between psychological flexibility and general mental health and physical well-being was stronger than the correlation between emotional intelligence and these outcome variables. In 2006, Bond and Flaxman found that job control and psychological flexibility predicted learning, performance, and mental health of 488 customer service center workers. In addition, there was an interaction effect of psychological flexibility on job control. Higher levels of psychological flexibility at Time 1 increased the beneficial effects of higher levels of job control on learning, performance and mental health. In a rehabilitation setting, McCracken and Yang (2008) surveyed 98 workers including nurses, physiotherapists, occupational therapists, physicians, speech and language therapists, psychologists and administrative staff. Psychological flexibility, mindfulness and values-based action were associated with less burnout, better health, and better well-being. Bond, Flaxman, and Bunce (2008) tested the extent to which a work reorganization intervention improved mental health, absence rates and job motivation, by enhancing perceived levels of job control. Participants who had higher levels of psychological flexibility perceived that they had greater levels of job control due to the intervention, which led to improvements in mental health and absence rates. Greater psychological flexibility thus had not only a direct positive effect on mental health, but also indirect benefits through enhanced job control.

It is clear from the research reviewed thus far that ACT interventions have been effective in a variety of organizations both in reducing psychological distress at work and in enabling workers to enhance their performance. Such interventions have demonstrated reliable positive effects on general mental health and decreases in depression, depersonalization, stigmatization and burnout. They have also been shown to empower employees to take on new work challenges and embrace positive change in work practices and structure. It is also clear that the constructs based on relational frame theory that are understood to underlie the efficacy of ACT interventions, in particular psychological flexibility, mediate the impact of ACT interventions in the workplace and are associated with increased employee resilience and performance improvement. In fact, Hayes et al. (2006) conducted a meta-analysis of 32 studies involving 6628 participants, across clinical and industrial/organizational psychology, to investigate the relationship between psychological flexibility and various quality of life outcomes and found a weighted effect size of these relations of .42. However, even though the greatest impact of relational frame theory on the workplace has undoubtedly been through ACT, researchers have recently attempted to employ RFT to investigate the effects of common organizational interventions and it is to that research that we now turn.

Analyses of Common Organizational Interventions

Within OBM, the majority of empirical studies have been motivated by and analyzed using direct contingency accounts, but, since 2001, a number of authors (Austin, 2001; Austin & Wilson, 2002; Hayes, 2004; Wiegand & Geller, 2004) have recommended including more complex accounts of human behavior, especially verbal behavior, to shed light on organizational behavior. This resulted in a special issue of the *Journal of Organizational Behavior Management* on potential contributions of RFT to OBM in 2006. Since then, two empirical studies have interpreted their data using conceptual accounts based on RFT, and these will be discussed here. The first of these focused on the interaction between rule-following and performance feedback and the second focused on goal setting.

Haas and Hayes (2006) investigated the effects of feedback on rule-governed behavior in a laboratory-based experiment. This study highlights the unpredictable effects of feedback on behavior and the intricate interrelationship between rules and feedback. The rules (networks of relational responses) that an individual uses to navigate the world and engage in work to some degree determine the effect of feedback on the individual's behavior. Haas and Hayes employed a complicated experimental paradigm in which participants moved a sign through a visual grid on a computer screen to earn points. In the first part of the experiment, participants were instructed on how best to earn points, but once participants had learned to do this, the computer game changed so that participants could earn more points by ignoring the original instructions. Participants were assigned to one of variety of rule conditions. One group received no instructions on how to best earn points during the first stage (Minimal Rule) and this served as the control group. Two of the other groups received accurate feedback on whether they were following the original rule ("You are/are not following the rule that you were given at the start of the session") and one of these also received performance feedback (e.g., "You earned 10 points"). Two more groups received random rule-following feedback (50% following, 50% not following regardless of performance) and one of these received performance feedback. During the experiment, the authors found that the Minimal Rule group were most sensitive to the change in experimental contingencies and the least sensitive group was the one that received accurate rule-following feedback and performance feedback.

It seems counter-intuitive that the group in Haas and Hayes' study who received the most accurate feedback would perform most insensitively. During the second part of the experiment, if participants in this group shifted their behavior to fit the new contingencies, they received messages telling them that they were not following the original instructions, but they also received performance feedback so they saw the impact of their behavior on the points earned. One might expect that participants would be more likely to ignore the rule-following feedback when they had access to performance feedback. However, the opposite was the case. Accurate performance feedback enhanced the effect of the rule-following feedback, which increased insensitivity and impaired performance. Haas and Hayes suggested that, when participants were told that they were "not following the rule" at the same time as they were told "You earned x points," this may have established the points as aversive if participants had a pre-experimental history of reinforcement for following rules and punishment for breaking rules. For these participants, points constituted evidence that they had not followed the rules provided by the experimenter. Such an account would explain the observed insensitivity effects.

Another common intervention in OBM that highlights the impact of verbal statements on the functions of feedback is goal setting. In the goal-setting literature, Locke, Shaw, Saari and Latham (1981) have argued that feedback is a critical determinant of goal-directed performance in that it allows the individual to gauge the relationship between his/her current performance and his/her goal. Behavioral researchers have made similar claims; both Agnew (1998) and O'Hora & Maglieri (2006) have suggested that goal setting works in part by establishing novel functions in performance feedback. According to the relational frame approach provided by O'Hora and Maglieri, feedback on current performance can serve to reinforce performance once goals have been set. Feedback can serve both positive ("I'm getting somewhere") and negative ("I still have work to do") reinforcing functions. For particular individuals in particular situations, the functions of performance feedback will be due to the individual's previous behavioral history in similar contexts (e.g., previous goal achievement).

Tammemagi, O'Hora and Maglieri (in press) investigated the impact of goal setting on the performance of 26 college undergraduates in an analog work setting. These researchers employed a data entry task designed to mimic a hospital task in which a technician might enter heart-rate data for various individuals into a database. This study incorporated two novel features. First, participants' performance was measured at baseline, which is not typical in the basic research on goal setting. Second, the paradigm included a choice condition, in which participants were asked to choose to work to a high or low goal. Participants were exposed to sequence of 5 phases: a baseline phase, then either a high or low goal condition (counterbalanced across participants), followed by a second baseline, the remaining goal condition and a final choice phase. As expected, performance increased significantly on introduction of a high specific goal. Mean performance increased by 21% when a high goal was introduced after baseline, whereas mean performance only increased by 11% when a low goal was the first goal condition. In the final phase, 20 of 26 participants chose the low goal condition (involving the least effort to attain), when given the choice between a low and a high goal condition.

O'Hora and Maglieri (2006) proposed that, even though higher goals will result in higher performance, achieving goal levels periodically is necessary in order to maintain performance. Unattainable goals may give rise to high levels of performance initially, but this level of behavior should decrease over time because reinforcement cannot be obtained. A high, unattainable goal may seem attainable over a short period of time, but over a longer duration, as the person's performance does not significantly progress his/her toward the goal, performance may decrease or cease altogether. In the Tammemagi et al. study, a negative trend in performance was observed within the high goal condition for a minority (38%) of participants. When the high goal was presented as the second goal condition, a negative trend was observed for more participants (46%) than when it was presented as the first condition (31%). Even though these findings constitute some evidence of extinction of goal-directed behavior, for many individuals, this behavior was quite resistant to extinction. Depending on an individual's behavioral history, prolonged exposure to unattainable goals may be necessary for extinction to occur.

Future Directions

Relational frame theory is beginning to make an impact on organizational behavior management. ACT interventions, based on RFT, have now demonstrated efficacy across a wide range of organizations in improving mental health outcomes and organizational performance. Correlational research on psychological flexibility has contributed to informing the picture of the modern worker by identifying the skills that workers need to avoid frustration and burnout and to positively affect change in the workplace. Conceptual analyses of common organizational interventions that have employed the concepts of RFT have suggested new research questions for organizational researchers, particularly dealing with how verbal stimuli affect employee behavior. The first studies to address these questions have been conducted but much remains to be done.

Recent RFT conceptual analyses of organizational phenomena suggest further areas of research in OBM. Stewart, Barnes-Holmes, Barnes-Holmes, Bond and Hayes (2006) provided an overview of Industrial-Organizational psychology that suggested numerous potential contributions of RFT in the workplace. Stewart et al. suggested RFT contributions in the areas of job satisfaction, attitudes, and behavior, teamwork, organizational culture and organizational development. For example, they suggest that attitudes be treated as relational networks that support long-term consistencies in behavior and that RFT-based interventions be used to modify attitudes or reduce the impact of attitudes on behavior if necessary. More recently, Herbst and Houmanfar (2009) provided an interpretation of organizational values informed by RFT that highlighted practical positive changes that can be made in the workplace.

While RFT-based translational research in OBM is in its infancy, ACT intervention research in organizational settings is now well established. The majority of this research has focused on the impact of ACT interventions on verbal dependent variables such as scores on mental health questionnaires. Particularly exciting from an OBM perspective, however, is the ACT research that has demonstrated improvements in observable behavior, such as reductions in errors and sick time and the adoption of new work practices. OBM practitioners and researchers might consider including ACT interventions in the suite of procedures that can be brought to bear on organizational issues. In addition, the positive effects of ACT interventions on workers' adjustment to novel work practices suggests that combining ACT workshops with traditional interventions such as employee training, goal setting, feedback, task clarification and process redesign may enhance the effect of these traditional interventions.

The foregoing suggestions constitute but a fraction of the myriad possibilities for new OBM research influenced by RFT. It is our hope that RFT will enable OBM researchers to begin to tease out the intricate details of behavioral control in the workplace. The concepts of RFT continue to show promise in enabling behavior analysts to address complex verbal behaviors within a comprehensive operant framework. In doing so, RFT broadens the scope of OBM interventions and provides copious opportunities for future translational and applied research in organizational settings.

References

Agnew, J. L. (1998). The establishing operation in organizational behavior management. Journal of Organizational Behavior Management, 18, 7-19.

Austin, J. (2001). Some thoughts on the field of Organizational Behavior Management. Journal of Organizational Behavior Management, 20(3-4), 191-202.

Austin, J., & Wilson, K. G. (2002). Response-response relationships in Organizational Behavior Management. *Journal of Organizational Behavior Management, 21*(4), 39-53.

Bond, F. W., & Bunce, D. (2000). Mediators of change in emotion-focused and problem-focused worksite stress management interventions. *Journal of Occupational Health Psychology, 5*(1), 156–163.

Bond, F. W., & Bunce, D. (2003). The role of acceptance and job control in mental health, job satisfaction, and work performance. *Journal of Applied Psychology, 88*(6), 1057–1067.

Bond, F. W., & Flaxman, P. E. (2006). The ability of psychological flexibility and job control to predict learning, job performance, and mental health. *Journal of Organizational Behavior Management, 26*(1-2), 113–130.

Bond, F. W., Flaxman, P. E., & Bunce, D. (2008). The influence of psychological flexibility on work redesign: Mediated moderation of a work reorganization intervention. *Journal of Applied Psychology, 93*(3), 645–654.

Bond, F. W., Flaxman, P. E., van Veldhoven, M. J. P. M., & Biron, M. (2010). The impact of psychological flexibility and acceptance and commitment therapy (ACT) on health and productivity at work. In J. Houdmont & S. Leka (Eds.), *Contemporary Occupational Health Psychology: Global Perspectives on Research and Practice* (pp. 296-313). Chichester, UK: Wiley-Blackwell.

Brinkborg, H., Michanek, J., Hesser, H., & Berglund, G. (2011). Acceptance and commitment therapy for the treatment of stress among social workers: A randomized controlled trial. *Behaviour Research and Therapy, 49*(6–7), 389–398.

Bucklin, B. R., Alvero, A. M., Dickinson, A. M., & Jackson, A. K. (2000). Industrial-Organizational Psychology and Organizational Behavior Management: An objective comparison. *Journal of Organizational Behavior Management, 20*(2), 27-75.

Dahl, J., Wilson, K. G., & Nilsson, A. (2004). Acceptance and commitment therapy and the treatment of persons at risk for long-term disability resulting from stress and pain symptoms: A preliminary randomized trial. *Behavior Therapy, 35*(4), 785–801.

Donaldson-Feilder, E. J., & Bond, F. W. (2004). The relative importance of psychological acceptance and emotional intelligence to workplace well-being. *British Journal of Guidance & Counselling, 32*(2), 187–203.

Flaxman, P. E., & Bond, F. W. (2010). A randomized worksite comparison of acceptance and commitment therapy and stress inoculation training. *Behaviour Research and Therapy, 48*(8), 816–820.

Ganster, D. C. (1989). *Measurement of worker control.* Final report to the National Institute of Occupational Safety and Health, Contract No. 88—79187

Haas, J. R., & Hayes, S. C. (2006). When knowing you are doing well hinders performance: Exploring the interaction between rules and feedback. *Journal of Organizational Behavior Management, 26*(1/2), 91-111.

Hayes, S. C. (1987). A contextual approach to therapeutic change. In N. Jacobson (Ed.) *Psychotherapists in clinical practice: Cognitive and behavioral perspectives.* (pp. 327–387). New York: Guilford Press.

Hayes, S. C. (2004). Fleeing from the elephant. *Journal of Organizational Behavior Management, 24*(1-2), 155–173.

Hayes, S. C, Barnes-Holmes, D., Roche, B., (Eds.). (2001). *Relational Frame Theory: A post-Skinnerian account of human language and cognition.* New York: Plenum.

Hayes, S. C., Bissett, R., Roget, N., Padilla, M., Kohlenberg, B. S., Fisher, G., Masuda, A., Pistorello, J., Rye, A. K., Berry, K., & Niccolls, R. (2004). The impact of acceptance and commitment training and multicultural training on the stigmatizing attitudes and professional burnout of substance abuse counselors. *Behavior Therapy, 35*(4), 821–835.

Hayes, S. C., Luoma, J. B., Bond, F. W., Masuda, A., & Lillis, J. (2006). Acceptance and Commitment Therapy: Model, processes and outcomes. *Behaviour Research and Therapy, 44*(1), 1–25.

Hayes, S. C., Masuda, A., Bissett, R., Luoma, J., & Guerrero, L. F. (2004). DBT, FAP, and ACT: How empirically oriented are the new behavior therapy technologies? *Behavior Therapy, 35,* 35–54.

Hayes, S. C., Strohsahl, K., Wilson, K. G., Bissett, R. T., Pistorello, J., Toarmino,… McCurry, S.M. (2004). Measuring experiential avoidance: A preliminary test of a working model. *The Psychological record, 54*(4), 553–578.

Hayes, S. C., Zettle R. D., & Rosenfarb, I. (1989). Rule following. In S. C. Hayes (Ed.) *Rule-governed behavior: Cognition, contingencies and instructional control* (pp. 191-220). New York: Plenum.

Herbst, S. A., & Houmanfar, R. (2009). Psychological approaches to values in organizations and Organizational Behavior Management. *Journal of Organizational Behavior Management, 29*(1), 47-68.

Lloyd, J., Bond, F. W., & Flaxman, P. E. (in press). Identifying psychological mechanisms underpinning a cognitive behavioral therapy intervention for emotional burnout. *Work & Stress.*

Locke, E. A., Shaw, K. N., Saari, L. M., & Latham, G. P. (1981). Goal setting and task performance: 1969-1980. *Psychological Bulletin, 90,* 125-152.

Luoma, J. B., Hayes, S. C., Twohig, M. P., Roget, N., Fisher, G., Padilla, M., Bissett, R., & Kohlenberg, B. (2007). Augmenting continuing education with psychologically focused group consultation: Effects on adoption of group drug counseling. *Psychotherapy: Theory, Research, Practice, Training, 44*(4), 463–469.

Marlatt, G. A., & Gordon, J. R. (1985). *Relapse Prevention: Maintenance strategies in the treatment of addictive behaviors.* New York: Guilford.

Maslach, C., Schaufeli, W. B., & Leiter, M. P. (2001). Job burnout. *Annual Review of Psychology, 52*(1), 397–422.

McCracken, L. M., & Yang, S.-Y. (2008). A contextual cognitive-behavioral analysis of rehabilitation workers' health and well-being: Influences of acceptance, mindfulness, and values-based action. *Rehabilitation Psychology, 53*(4), 479–485.

O'Hora, D., & Maglieri, K. A. (2006). Goal statements and goal-directed behavior: A relational frame account of goal setting in organizations. *Journal of Organizational Behavior Management, 26*(1/2), 131–170.

Ruiz, C. O., Rios, F. L., & Martin, S. G. (2008). Psychological intervention for professional burnout in the palliative care unit at Gregorio Maranon University Hospital. *Medicina Preventiva, 15*(2), 93-97.

Stewart, I., Barnes-Holmes, D., Barnes-Holmes, Y., Bond, F. W., & Hayes, S. C. (2006). Relational frame theory and industrial/organizational psychology. *Journal of Organizational Behavior Management, 26*(1/2), 55–90.

Tammemagi, T., O'Hora, D., & Maglieri, K. A. (in press). The effects of a goal setting intervention on productivity and persistence in an analogue work task. *Journal of Organizational Behavior Management.*

Varra, A. A., Hayes, S. C., Roget, N., & Fisher, G. (2008). A randomized control trial examining the effect of acceptance and commitment training on clinician willingness to use evidence-based pharmacotherapy. *Journal of Consulting and Clinical Psychology, 76*(3), 449–458.

Wiegand, D. M., & Geller, E. S. (2004). Connecting Positive Psychology and Organizational Behavior Management. *Journal of Organizational Behavior Management, 24*(1-2), 3–25.

Simon Dymond, PhD, BCBA-D, is a reader in psychology at Swansea University. He received his undergraduate training and PhD (in 1996) from University College Cork, where he studied under Dermot Barnes-Holmes. He has published over seventy empirical research articles on derived relational responding, avoidance, and gambling, and currently sits on several editorial boards of publications, including the *Journal of the Experimental Analysis of Behavior* and *The Psychological Record.*

Bryan Roche, PhD, CPsychol, CSci, AFBPsS, graduated with his doctorate in behavior analysis in 1995, after which he took up academic posts at University College Cork, Ireland and the University of Bath, UK. His current position is at the National University of Ireland, Maynooth. Roche has published approximately eighty articles, peer-reviewed papers, and book chapters on relational frame theory (RFT) and related topics. In particular, his research has involved the application of RFT to the study of social and sexual behavior, the understanding and treatment of anxiety, and most recently the development of online relational frame training interventions to increase intelligence quotients (raiseyouriq. com). He was coeditor of the book *Relational Frame Theory: A Post-Skinnerian Analysis of Language and Cognition* (2001). Roche currently sits on the editorial boards of several behavior-analytic journals, and is a regular ad-hoc reviewer for several of the major international journals of behavioral psychology.

Foreword writer **Jan De Houwer, PhD**, has authored and coauthored more than 160 publications in international journals, including *Psychological Bulletin*; *Journal of Experimental Psychology: General*; and *Behavioral and Brain Sciences*. He is currently editor of the journal *Cognition and Emotion.*

Index

I

identity matching, 56–57

Implicit Association Test (IAT), 99, 110, 111, 112

implicit cognition, 97–118; cognitive approach to, 98–100; functional approach to, 101; IRAP procedure and, 104–118; REC model and, 102–104

implicit evaluations, 98, 103

implicit goals, 41

Implicit Relational Assessment Procedure (IRAP), 104–114; challenges and future directions for, 114–118; experiential avoidance study using, 239; methodology and uses for, 104–109; ongoing refinements to, 112–114; research on, 109–112, 116–117, 181

indirect procedures, 99–100

Innovation Promotion Program (IPP), 255

intermittent schedules of reinforcement, 53

interrupted chain procedure, 156, 157

intraverbals: definition of, 163–164; derived intraverbals, 163–166

IQ measurement, 140, 184, 188, 192

IRAP. See Implicit Relational Assessment Procedure

I-YOU relations, 129–130, 135, 136, 137, 142

J

James, William, 6

job control, 259, 260

Job Control Questionnaire, 259

Joe the Bum metaphor, 228

Journal of Organizational Behavior Management, 261

K

Kaufman Brief Intelligence Test (K-BIT), 181

kinship relations, 133

L

labeling, 161

language: behavior analytic conception of, 52; developmental explosion of, 66; human evolution of, 16; relational framing and, 65–67, 152–155; second language learning and, 167–168; stimulus equivalence and, 55, 61

language instruction, 151–173; derived intraverbals and, 163–166; derived mands and, 156–161; derived tacts and, 161–163; future directions in, 172–173; perspective-taking and, 168–172; RFT approach to, 152–155; second language learning and, 167–168; verbal operants and, 155–156

Less-than relations, 80, 87

lexical decision task, 77

Long, Douglas M., 5

M

Maglieri, Kristen, 253

mands: definitions of, 156, 179; derived mands, 156–161; reversed mands, 161

P

Palca, Joe, 39

peer review, 38

Pepper, Stephen, 6, 29

performance feedback, 262, 263

perspective relations: cognitive abilities related to, 183–184; multiple exemplar training of, 189

perspective-taking: ACT processes based on, 225–226; deictic relations and, 168–172, 183–184; RFT conceptualization of, 129–130, 183–184; theory of mind approach to, 128–129. *See also* relational perspective-taking

phobias, 207–208

post-experimental reports, 81

post-modernism, 17

pragmatic truth criterion, 45

present moment awareness, 225, 240

priming effects, 77–78, 83

protocol analysis, 81, 91

psychological flexibility: ACT definition of, 224–225; RFT perspective on, 240–243, 258–259; workplace interventions and, 258–261

psychopathology, 199–212; cognitive defusion and, 208–211; derived fear and avoidance and, 202–206; neuroimaging studies of, 212; RFT view of, 200–202; specific phobias and, 207–208

public goals, 41–43

Putnam, Hilary, 28, 30–31

R

radical behaviorism, 20, 220

reaction time (RT) measures, 77–81

realism: explanation of, 17–18; functional approach and, 19–20

received view of science (RVS), 28–29

reflexivity, 54–55, 60

Rehfeldt, Ruth Anne, 151

reinforcement: differential, 171; intermittent schedules of, 53; nonspecific, 161, 164

reinforcer establishing effect, 156

Relapse Prevention Model, 257

relational coherence, 103–104

Relational Completion Procedure (RCP), 75–76

Relational Elaboration and Coherence (REC) model, 102–104

Relational Evaluation Procedure (REP), 75

relational flexibility, 114

Relational Frame Theory (RFT): ACT linked to, 237–243; background to, 52–56; contextual behavioral science and, xi, 8; contributions to cognitive science, xi–xii; cooperative communication and, 14–16; DRR research and, 73–93; evolution science and, 21; implicit cognition and, 101, 118; language and, 65–67, 152–155; organizational behavior and, 254–255, 258–259, 261–265; overview of, 51–67; perspective-taking in, 129–130, 183–184,

Wechsler Adult Intelligence Scale (WAIS-III), 181
Whelan, Robert, 73
Whiteman, Kerry, 27
Wilson, Kelly G., 27
word-object relations, 132
workplace: ACT interventions in, 255–261, 265; OBM interventions in, 261–264; organizational behavior in, 253–255; psychological flexibility in, 258–261; RFT applied to, 254–255, 258–259, 261–265
worldviews, 29–30

XYZ

Zamore, Phillip, 39

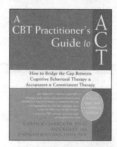

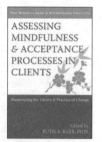
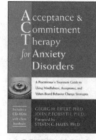